Springer Series in Applied Biology

Immobilised Macromolecules: Application Potentials

Springer Series in Applied Biology

Series Editor: Prof. Anthony W. Robards PhD, DSc, FIBiol

Published titles:

Foams: Physics, Chemistry and Structure

The 4-Quinolones: Antibacterial Agents in Vitro

Food Freezing: Today and Tomorrow

Biodegradation: Natural and Synthetic Materials

Proposed future titles:

Probiotics: Bacterial Prophylaxis
Ed. S. A. W. Gibson

Immobilised Macromolecules:
Application Potentials

Edited by U. B Sleytr, P. Messner, D. Pum and M. Sára

Springer-Verlag London Ltd.

U. B. Sleytr, P. Messner, D. Pum and M. Sára
Zentrum für Ultrastrukturforschung, Universität für Bodenkultur,
A-1180 Vienna, Austria

Series Editor
Professor Anthony William Robards, BSc, PhD, DSc, DipRMS,
FIBiol
Director, Institute for Applied Biology, Department of Biology,
University of York, York YO1 5DD, UK

Cover Illustration: Electron micrograph showing the regular
pattern of polycationised ferritin molecules attached to the S-layer
of an archaebacterium

ISBN 978-1-4471-3481-7   ISBN 978-1-4471-3479-4 (eBook)
DOI 10.1007/978-1-4471-3479-4
British Library Cataloguing in Publication Data
Immobilised Macromolecules:Application Potentials. - (Springer Series in Applied
Biology)
I. Sleytr, U. B. II Series
574. 19

Library of Congress Cataloging-in-Publication Data
Immobilised macromolecules:application potentials / edited by U. B. Sleytr...(et al.)
p. cm. - (Springer series in applied biology)
Includes index.

1. Immobilized proteins. 2. Immobilized enzymes. 3. Immobilized ligands (Biochemistry)
I. Sleytr, U. B. (Uwe Bernd) II. Series.
TP248.65.I47I55 1992
660'.63 - dc20   92-3291
                                                               CIP

© Springer-Verlag London 1993
Originally published by Springer-Verlag Berlin Heidelberg New York in 1993
Softcover reprint of the hardcover 1st edition 1993

Set by Institute for Applied Biology, Department of Biology, University of York
12/3830 - 543210 Printed on acid free paper

Foreword from Series Editor

The Institute for Applied Biology was established by the Department of Biology at the University of York to consolidate and expand its existing activities in the field of applied biology. The Department of Biology at York contains a number of individual centres and groups specialising in particular areas of applied research which are associated with the Institute in providing a comprehensive facility for applied biology. Springer-Verlag has a long and successful history of publishing in the biological sciences. The combination of these two forces leads to the "Springer Series in Applied Biology". The choice of subjects for seminars is made by our own editorial board and external sources who have identified the need for a particular topic to be addressed.

The first volume, *"Foams: Physics, Chemistry and Structure"*, has been quickly followed by *"The 4-Quinolones: Antibacterial Agents in Vitro"*, *"Food Freezing: Today and Tomorrow"* and *"Biodegradation: Natural and Synthetic Materials"*. The aim is to keep abreast of topics that have a special applied, and contemporary, interest. The current volume describes how biomolecules have far reaching uses relating to the development of biosensors, optrodes, conjugated vaccines, biocompatible surfaces and membranes with novel properties.

Up to three volumes are published each year through the editorial office of *IFAB Communications* in York. Using modern methods of manuscript assembly, this streamlines the publication process without losing quality and, crucially, allows the books to take their place in the shops within four to five months of the actual seminar. In this way authors are able to publish their most up-to-date work without fear that it will, as so often happens, become outdated during an overlong period between submission and publication.

The applications of Biology are fundamental to the continuing welfare of all people, whether by protecting their environment or by ensuring the health of their bodies. The objective of this series is to become an important means of disseminating the most up-to-date information in this field.

York, July 1992   A. W. Robards

Editor's Preface

The last few years have seen astonishing progress in the fields of biotechnology, bioengineering and biomedical engineering. In particular, there have been many and far-reaching applications relating to the development of biosensors, optrodes, conjugated vaccines, biocompatible surfaces and membranes with novel properties. Common to all these interdisciplinary areas of research is a requirement for the immobilisation of functional macromolecules which, in turn, highlights the importance of knowing much more about how biomolecules interact and how they can be engineered to produce functionally useful surfaces and membranes. The contents of this book point to the way that scientists from many different disciplines are now coming together to bring their expertise to bear on the development of novel technology that can only arise from collaboration between protein- and physical chemists, enzymologists, biotechnologists, microbiologists, optical physicists and many others. The potential developments from such synergistic relationships are truly awe-inspiring whether they involve "clever" membranes that can be used to filter and/or adsorb molecules with a very high degree of specificity, the production of a wider range of biosensors, or opening the way for new vaccines. All such work needs to be underpinned by research on the molecular building blocks of the new devices and many of the relevant studies are reported in this book. The different chapters range across topics including: protein folding, stability and assembly; orientated immobilisation of antibodies; avidin/biotin interactions; 2D-protein crystals; and enzyme hapten or antigen immobilisation. From these fundamental studies it has been possible to lead on to discuss practical applications including; affinity and ultrafiltration membranes; biosensors; and diagnostics. In consequence, this book provides a unique and up-to-the-minute account of the state-of-the-art in the different ways that immobilised macromolecules can be used in practical applications.

U. B. Sleytr

Contents

Contributors

Dr. E. A. Bayer
Department of Biophysics, The Weizmann Institute of Science, 76100 Rehovot, Israel

Dr. U. Bilitewski
GBF Braunschweig-Stöckingen, Abteilung Enzymtechnologie, Mascheroder Weg 1, D-3300 Braunschweig-Stöckheim, Bundesrepublik Deutschland

Dr. S. Birnbaum
Chemical Center, University of Lund, S-221 00 Lund, Sweden

Professor J. M. Courtney
Bioengineering Unit, University of Strathclyde, Wolfson Centre,
106 Rottenrow, Glasgow G4 0NW, Scotland

Dr. P. W. Feldhoff
Medical Oncology and Hematology, James Graham Brown Cancer Center, University of Louisville School of Medicine, Louisville KY 40292, USA

Mr. G. Hawa
Institut für Allgemeine Biochemie, Universität Wien, Währingerstraße 38, 1090, Wien, Austria

Professor Dr. R. Jaenicke
Institut für Biophysik und Physikalische Biochemie, Universität Regensburg W-8400 Regensburg, FRG

Professor J. Kas
Department of Biochemistry and Microbiology, Institute of Chemical Technology, 166 28 Prague 6, Czechoslovakia

Professor G. Kopperschläger
Institute of Biochemistry, Medical Faculty, University of Leipzig, FRG

Dr. J. Kirchberger
Institute of Biochemistry, Medical Faculty, University of Leipzig, FRG

Dr. S. Küpcü
Zentrum für Ultrastrukturforschung und Ludwig Boltzmann Institut für
Molekulare Nanotechnologie, Universität für Bodenkultur, A-1180 Vienna,
Austria

Dr. A. J. Malcolm
Alberta Research Council, Box 8330, Station "F", Edmonton, Alberta,
T6H 5X2 Canada

Dr. E. Mann-Buxbaum
Institut für Allgemeine Biochemie, Universität Wien, Währingerstraße 38,
1090, Wien, Austria

Dr. P. Messner
Zentrum für Ultrastrukturforschung, Universität für Bodenkultur, A-1180
Vienna, Austria

Ms. I. Moser
Institut für Allgemeine Biochemie, Universität Wien, Währingerstraße 38,
A-1090, Wien, Austria

Professor Dr. F. Pittner
Institut für Allgemeine Biochemie, Universität Wien, Währingerstraße 38,
A-1090, Wien, Austria

Dr. D.Pum
Zentrum für Ultrastrukturforschung, Universität für Bodenkultur, A-1180
Vienna, Austria

Dr. J. Sajdok
Department of Biochemistry and Microbiology, Institute of Chemical
Technology, 166 28 Prague 6, Czechoslovakia

Dr. M. Sára
Zentrum für Ultrastrukturforschung, Universität für Bodenkultur, A-1180
Vienna, Austria

Dr. Th. Schalkhammer
Institut für Allgemeine Biochemie, Universität Wien, Währingerstraße 38,
1090, Wien, Austria

Professor Dr. U. B. Sleytr
Zentrum für Ultrastrukturforschung und Ludwig Boltzmann Institut für
Molekulare Nanotechnologie, Universität für Bodenkultur, A-1180 Vienna,
Austria

Dr. R. H. Smith
1010 Buchanan Place, Edmonton, Alberta T6R 2A6 Canada

Dr. F. Strejček
Department of Biochemistry and Microbiology, Institute of Chemical
Technology, 166 28 Prague 6, Czechoslovakia

Dr. S. Sundaram
Bioengineering Unit, University of Strathclyde, Wolfson Centre,
106 Rottenrow, Glasgow G4 0NW, Scotland

Dr. F. M. Unger
Alberta Research Council, Box 8330, Station "F", Edmonton, Alberta,
T6H 5X2 Canada

Dr. G. Urban
Institut für Allgemeine Biochemie, Universität Wien, Währingerstraße 38,
1090, Wien, Austria

Mr. C. Weiner
Zentrum für Ultrastrukturforschung und Ludwig Boltzmann Institut für
Molekulare Nanotechnologie, Universität für Bodenkultur, A-1180 Vienna,
Austria

Mr. S. Weigert
Zentrum für Ultrastrukturforschung und Ludwig Boltzmann Institut für
Molekulare Nanotechnologie, Universität für Bodenkultur, A-1180 Vienna,
Austria

Professor M. Wilcheck
Department of Biophysics, The Weizmann Institute of Science, 76100
Rehovot, Israel

Professor Dr. O. S. Wolfbeis
Institut für Organische Chemie, Analytische Abteilung, Karl-Franzens-
Universität Graz, Heinrichstraße 28, A-8010 Graz, Austria

Mrs. J. Yu
Bioengineering Unit, University of Strathclyde, Wolfson Centre,
106 Rottenrow, Glasgow G4 0NW, Scotland

Chapter 1

Protein Stability, Folding and Association

R. Jaenicke

Introduction

Significant advances have been made in recent years in the technology of protein reconstitution. Making use of the stunning fact that even giant protein complexes such as ribosomes or multienzyme systems, or kringle structures like tissue plasminogen activator can be subjected to denaturation/renaturation, recombinant proteins are now "renativated" on large scale. The basis of this kind of industrial application is still alchemy, for two reasons: first, neither the mode nor the code of protein folding has been solved yet; and, second, protein stability is still not well understood in its thermodynamic principles.

Starting points for the present discussion are native mesophilic globular proteins of average molecular mass. From the available data base with ca. 500 crystal structures it has become clear that beyond a critical length of the polypeptide chain of ≤ 150 amino acid residues, proteins consist of domains. They represent cooperative units with respect to both intrinsic stability and folding. The general principle underlying the three-dimensional structure of proteins in aqueous solution is the minimisation of hydrophobic surface area which leads to the well-known distribution of non-polar residues in the interior of the molecule, and polar residues on its surface. A given amino acid composition does not necessarily allow this "micellar structure" to be perfect so that polar residues may be buried in the interior and non-polar residues may be exposed to the solvent. Evidently, the distribution of residues and the packing of the core has important implications for protein dynamics and stability. Both are determined by the subtle balance of attractive and repulsive interactions between the residues on one hand and the solvent on the other (Jaenicke 1987a).

Protein Stability

Interactions Involved in Protein Stability

The contributions of the various weak interactions that determine the three-dimensional structure of proteins are more controversial than they used to be since Kauzmann wrote his classical review (Kauzmann 1959; Privalov and Gill 1989; Dill 1990; Creighton 1991; Jaenicke 1991a; Jaenicke 1991b). In the present context, it is sufficient to note that globular proteins in solution exhibit only marginal free energies of stabilisation. The subtle balance of hydrogen bonds, hydrophobic interactions, ion pairs and van der Waals forces yields values for the free energy of stabilisation, ΔG_{stab}, which are generally found to be below 60 kJ/mol, independent of the mode of denaturation (Pace 1990; Pfeil 1986). Expressed in ΔG_{stab} per residue, this means that the free energy is one order of magnitude below the thermal energy (kT). Thus, on balance, the stability of globular proteins in solution is based on the equivalent of a few weak interactions. This holds even in the case of highly stable proteins, e.g. from hyperthermophilic microorganisms (Jaenicke 1991b, 1992).

Ribonuclease T1 (RNaseT1) may serve as an example: Its structure is stabilised by two disulfide bridges, apart from 87 intramolecular hydrogen bonds and prominent hydrophobic interactions involving 85% of the non-polar residues; however, the free energy of stabilisation, ΔG_{stab}, does not exceed 24 kJ/mol (pH 7.0, 25°C). The high-resolution X-ray analysis of the protein has shown that the hydrophobic core of the molecule is sandwiched between a long α-helix and an extended antiparallel ß-sheet. A number of mutants with single amino-acid exchanges have been investigated, indicating that the stability of the molecule can be increased or decreased when residues in the helix or in the ß-sheet gain or lose a charge. In addition, substitution of side chains involved in intramolecular hydrogen bonds has proven that there is a net loss in stability per H-bond of the order of 5 kJ/mol, settling the old controversy whether there is a difference in H-bond strength between water-water and water-protein hydrogen bonds. Making up the balance of the various intra- and intermolecular interactions involved in the total free energy of stabilisation, it becomes clear that H-bonding, at least in small proteins, contributes equally, or even to a higher extent, to the configurational stability than hydrophobic interactions. The following numbers illustrate this conclusion: 494 kJ/mol for the conformational entropy, as compared to 268 kJ/mol for hydrophobic interactions and 448 kJ/mol for hydrogen bonds (Pace *et al. 1991;* Shirley *et al.* in press).

Coulomb interactions are well understood in model systems, but they become highly complex in non-homogeneous environments such as folded proteins, mainly because of the ill-defined dielectric constant in the immediate surrounding of the charges (Dill 1990; Sharp and Honig 1990; Creighton 1991). Whether or not ion-pairs contribute significantly to protein stability has been questioned since it became clear that most charged groups in globular proteins are exposed to the aqueous solvent (Kauzmann 1959). On the average, only one ion-pair per 150 amino acid residues of a globular protein is buried within the interior core (Barlow and Thornton 1983). Thus, only surface ion-pairs are expected to be involved in stabilisation. As indicated by site-directed mutagenesis and X-ray analysis, their contribution may be significant to the extent that thermophilic adaptation has been attributed to charge interactions (Perutz and Raidt 1975; Anderson *et al.* 1990). However, even for these groups, effects of varying salt concentration suggest that a significant fraction of the electrostatic free energy arises from the entropy of proton and water release rather than charge energetics (Stigter and Dill 1990). On the other hand, the destabilisation of proteins at extremes of pH has always been explained in terms of increased repulsive interaction at high net charge. Whether this is correct, or whether the ionisation of buried polar residues is involved is an open question (Privalov and Gill 1989; Pace *et al.* 1990).

An important observation in this context is the fact that at very low pH, charge effects seem to be inverted such that the acid denatured protein undergoes a second structural transition into its "A-state" or other "alternative states" (Goto *et al.* 1990; Buchner *et al.* 1991). These show structures distinct from both the native and denatured states, and are frequently identified as "molten globules" (see below); in certain cases they obviously differ from this rather ill-defined intermediate "collapsed state" by showing well-defined, but non-native structure (Buchner *et al.* 1991). The simplest explanation for the regain of spatial organisation of the polypeptide chain after preceding acid denaturation would be the fact that at pH $\leq$ 1 the high activity of the acid leads to a significant increase in ionic strength. This will screen part of the charge effects, thus modulating coulombic interactions. On the other hand, effects on water structure as well as preferential salt binding might be involved. These phenomena are expected to be relevant especially for peripheral proteins from acidophilic microorganisms.

The mechanisms underlying the stabilisation of proteins that undergo no or relatively slow turnover, such as basic pancreatic trypsin inhibitor (BPTI) or γ-crystallin, may be totally different from those discussed so far. The fact that reduction of the three cystine bridges in BPTI leads to complete unfolding, even in the absence of denaturants, indicates that in this case evidently folding and stability are coupled to disulfide bond formation (Creighton 1978, 1988a; Wetzel *et al.* 1990). Calorimetric studies on BPTI analogs, selectively modified with respect to the number of disulfide bonds, clearly show the gradual stabilisation of the protein with increasing crosslinking. In some cases, the stabilising effect is much larger than one would expect for the restricted increase in entropy of the denatured state. Since γ-crystallin does not contain disulfide bridges, in this case the low turnover must be related with the all-ß structure; a clear-cut explanation for the anomalous stability of the protein (at pH 1-10 and temperatures up to 75°C, or in the presence of 7M urea) has yet to be given (Rudolph *et al.* 1990).

The oxidation of cysteine residues to form cystine bridges is a cotranslational process which is speeded up by its "vectorial" character (Bergman and Kuehl 1979); in addition, the reaction is catalysed by protein disulfide isomerase (PDI). The enzyme resides in the endoplasmic reticulum (ER), where folding of secretory disulfide-bonded proteins is known to occur (Freedman 1984). It is a dimer with two internally homologous domains in each monomer which are structurally and functionally related with thioredoxin. Mechanistically the two proteins are linked as a redox pair with PDI as the weaker reductant; this property may be significant in ensuring that PDI does not reduce correct disulfide bonds that are already stabilised in the nascent native-like protein (Freedman 1989; Schmid 1991).

Structural Elements, Fragments, Domains and Subunits

Up to this point, intact proteins have been considered, with the general conclusion that stability is accomplished by the cumulative effect of non-covalent interactions at many locations within the globular entity of the protein molecule. In considering the stability of local structural elements, one may ask what happens if we shorten the protein, e.g. by limited proteolysis, and what is the minimum length of a polypeptide chain that is required to still form an intrinsically stable native-like structure (Jaenicke 1987a, 1991a; Baldwin 1990). NMR data have shown that oligopeptides down to six residues do form stable (non-random) conformations, supporting the idea that local structures may serve as "seeds" in the folding process (Wright *et al.* 1988). However, there is some indication that the short fragments have no substantial tendency to adopt the same conformation in unrelated protein structures; for example, reverse turn motifs observed in small peptides seem to be absent in the known three-dimensional structures of proteins with these sequences (Creighton 1988b). With regard to the stability of protein fragments, it has been known for a long time that protein domains show high intrinsic stablity, not far from the free energies observed for the uncleaved parent molecule (Jaenicke 1987a). Reducing the chainlength further, it becomes evident that proteins are cooperative structures showing mutual stabilisation of structural elements. In order to find out at which fragment size native-like structure can no longer be formed, thermolysin was used as a model (Table 1.1). Folding/unfolding experiments with a variety of BrCN fragments show that the N-terminal portion of the enzyme stabilises the all-helical C-terminal domain. This may be shortened drastically, down to the 62-residues three-helix structure, without aggregating or losing much of the stability of native thermolysin; only the C-terminal 20-residue helix is too short to maintain its native structure in aqueous solution (Vita *et al.* 1989). Whether the N- and the C-terminal ends of the polypeptide chain are important for protein stability depends on the protein. Taking RNase and lactate dehydrogenase (LDH) as examples, it has been shown that the N-terminal ends of both proteins can be cleaved without altering the overall topology; however, the stability is greatly affected. Cleaving off the C-terminus of RNase is sufficient to block the oxidative reshuffling reaction (Opitz *et al.* 1987; Teschner and Rudolph 1989).

Table 1.1. Physicochemical characteristics of thermolysin and its fragments: M, molecular mass; helicity from CD; ΔG_{stab}, free energy of stabilisation

Sequence	M(calc)	M(obs)	Helicity	T (denaturation)	ΔG_{stab}
1-316	34 227	34 800	100%	87°C	55 kJ/mol
121-316	20 904	23 000	96	74	47
206-316	11 829	11 900	95	67	31
225-316	9 560	9 000	92	65	26
255-316	6 630	≤12 000	100±10	64	20

As mentioned in connection with domain proteins, cooperative effects contribute significantly to the intrinsic stability of proteins. This feature is even more pronounced in protein assemblies.

To illustrate this, lactate dehydrogenase may again serve as an example. The *native tetramer* represents a dimer of dimers. Medium denaturant concentrations, as well as cleaving off the N-terminal decapeptide allow the *dimer* to be characterised: its stability is drastically reduced, it shows no enzymatic activity unless structure-making salts are added, and its folding is strongly impaired by irreversible side reactions. The *monomer* is only accessible as short-lived intermediate on the pathway of folding and association: limited proteolysis indicates a still higher degree of flexibility compared with the dimer, and no residual enzymatic activity can be observed. The domains (after "nicking") are able to pair correctly: they recognise each other in a topologically correct manner, so that one may assume that under quasiphysiological conditions "structured monomers" possess a native-like tertiary structure, but dramatically reduced stability (Jaenicke 1987a).

Physiological Stress and Protein Stability

Since it is not trivial to design proteins with specific stability characteristics, it seems useful to try to elucidate the long-term experiment that has been going on in nature, where adaptation to extremes of physical conditions in the biosphere has led to proteins with highly specified stabilities toward temperature, pressure, pH and low water activity. As taken from more than one century of careful investigation, life on earth is ubiquitous, except for centers of volcanic activity. The biosphere (including zones where "cryptobiosis" prevails) refers to the oceans with pressures up to 120 MPa (1200 atmospheres), the soil to a depth of 10-20 meters, and the atmosphere and stratosphere. Thus, organisms had to adapt their cell inventory to the extremes summarised in Table 1.2.

Table 1.2. Limits of viability and extremophilic adaptation[a]

Temperature	-5 - 110°C	Psychro- and thermophily
Pressure	0.1 - 120 MPa	Barophily
pH	0 - 12	Acido- and alkalophily
Water activity	0.6 - 1.0	Halophily

[a] Short-term survival and cryptobiosis may exceed the given limits

In comparing mesophiles and extremophiles, it has been shown that the range of viability is commonly shifted rather than broadened (Jaenicke 1981). Focusing on homologous proteins from organisms taken from different extreme environments, it turns out that adaptation to extremes of physical conditions tends to maintain "corresponding states" regarding overall structure, flexibility and ligand affinity. Generally, evolution is geared to maintain optimum function in widely differing solvent environments. It does so with an amazingly high degree of conservatism with respect to the three-dimensional "topology" of proteins and their constituent amino acids. This is beautifully illustrated by the fact that hemoglobin has preserved its characteristic peptide fold in spite of only two out of ca. 150 residues along the polypeptide chain having been absolutely conserved over the whole time span of evolution; on the other hand, all proteins, including those from the most extremophilic organisms, do not contain other than the 20 natural amino acids as amide-linked polypeptides. The increase in overall stability is of the same order as mentioned before: $\Delta\Delta G_{stab}$ for a typical thermophilic protein is $\approx$ 50 kJ/mol, again only marginal compared with the inner energy which is of the order of 10^4 kJ/mol (Baldwin and Eisenberg 1987). Faced with this ratio, it is evident that there may be many ways to accumulate the $\Delta\Delta G_{stab}$ required to stabilise a protein; on the other hand, one would predict that it is highly improbable to uncover general strategies of molecular adaptation. In responding to extreme physical conditions, organisms may use strategies other than molecular adaptation: acidophiles or alkalophiles avoid extremes of pH in their cytoplasm by pumping protons; halotolerant organisms may compensate salt gradients through isosmotic concentrations of "compatible" cytoplasmic compounds; thermotolerance may be brought about by extrinsic effectors such as cyclic 2.3 diphosphoglycerate etc. (Hensel and König 1988; Huber *et al.* 1989; Jaenicke 1991b) . The well-established stabilising effect of additives such as salts or polyols has recently been elucidated in terms of preferential solvation (Timasheff and Arakawa 1989); it will be discussed in connection with the protein folding problem.

Solvent Effects on Weak Interactions

As one would predict from the inverse solubility properties of polar and non-polar compounds in water (Privalov and Gill 1989), and from the fact that average proteins contain about the same amount of hydrophilic and hydrophobic residues, the influence of temperature on protein stability is highly complex, and still not fully understood. Due to the parabolic shape of the temperature profile of ΔG_{stab} (Privalov and Gill 1989), one would postulate two denaturation transitions:

i   endothermic (entropy-driven) "heat denaturation" at high temperature, and
ii exothermic (enthalpy-driven) "cold denaturation" at low temperature.

Both have been recently reviewed (Ghélis and Yon 1982; Privalov 1990; Jaenicke 1990). Therefore, it is sufficient to briefly summarise the effect of the relevant physical solvent parameters on the various weak interactions.

As taken from solubility studies, hydrophobic and Coulomb interactions become weaker with decreasing temperature, based on their entropic origin on one hand, and the temperature dependence of the dielectric constant, on the other. Hydrogen bonds are favored at low temperature and become weaker as temperature is increased. High electrolyte concentrations (simulating the extreme salinity in the case of halophily) are expected to strengthen hydrophobic interactions. Whether at extreme salt concentrations charge shielding affects Coulomb interactions significantly, or whether effects on water structure dominate the intermolecular weak forces in halophiles, is unresolved. The observation that specific "compatible ions" are required for full biological function supports the second alternative. It is self-evident that in the special case of low and high pH, it is mainly electrostatic repulsion and the exposure of buried titratable groups with anomalous pK-values that dominate the weak intermolecular interactions, apart from the high ion concentration (Kauzmann 1959; Jaenicke 1987a; Dill 1990; Goto *et al.* 1990; Creighton 1991).

In the case of high hydrostatic pressure, Braun-Le Chatelier's principle governs possible effects on protein stability. Volume changes during reactions in aqueous media are largely due to changes in solvent structure caused by solute-solvent interactions. Model studies on simple chemical reactions (protonation/deprotonation, isomerisation, hydrolysis, association, etc.), as well as complex biochemical reactions (including bacterial growth) allow to predict which processes might be crucial in high-pressure adaptation (Jaenicke 1981; Jaenicke 1987b). Using the size and sign of the reaction volumes as criteria, hydrophobic hydration and dissociation of ions are expected to be promoted at elevated pressure. As a consequence, disassembly of proteins or protein complexes and pH variations are observed (Fig. 1.1). Since the maximal pH shift over the whole ecologically relevant pressure range does not exceed ≈ 0.3 pH units, the latter effect does not seem to be of biological importance. Formation or breakage of H-bonds has no significant volume effect (Jaenicke 1981).

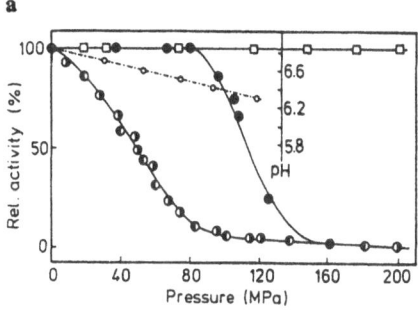 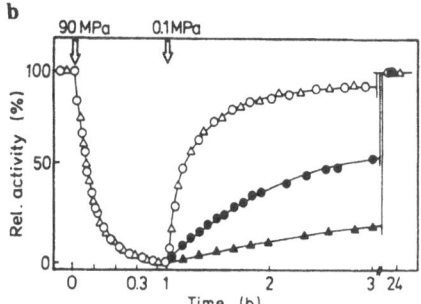

Fig. 1.1. Pressure-dependent dissociation of lactate dehydrogenase (pig heart) c = 25 μg/ml in Tris pH 7.6, I = 0.16M, 20°C. (a) Pressure-dependent equilibrium transition.(◐,◑) forward and backward reaction for the apoenzyme; (●,▫) holoenzyme (>94% NAD⁺) in the absence and in the presence of 0.2 M phosphate: (o) pH shift in 0.1 M phosphate buffer. (b) Kinetics of denaturation at 90 MPa (900 atm) and renaturation at atmospheric pressure. Activity (o) and intrinsic fluorescence (Δ) at 345 nm (λ_{exc} = 280 nm); enzyme concentrations: 720 nM (o), 42 nM (●) and 21 nM (▲) (Schade *et al.* 1980).

In discussing the effects of solvent variables on the forces stabilising biopolymers, two questions are of major importance:

i   Are the well-defined physiological and environmental conditions that are characteristic for a given organism in its specific biotope a necessary requirement to form and maintain the functional structure of its proteins, and

ii   What determines the definite limit of growth and reproduction outside the ranges of temperature, pressure, pH and salinity given in Table 1.2.

i The subtle balance of attractive and repulsive forces determining the native structure of proteins would suggest that the expression and folding of proteins in their active form require exactly defined conditions characteristic for the respective organism under optimum physiological conditions. However, cloning and expression of genes from extremophiles in mesophilic hosts have clearly been shown to result in native extremophilic proteins in active form. Thus, e.g. natural thermophilic conditions are obviously no necessary requirement for the correct folding of intrinsically thermostable proteins. In the case of proteins from halophilic organisms the guest-host relationship seems to be more critical. As will be shown, *in vitro* folding and reassociation frequently require careful optimisation of the solvent conditions; often "unphysiological conditions" do not allow reconstitution. Obviously, in this case the stability of intermediates is involved, rather than the stability of the native protein in its fully folded state. Whether in the cell specific factors, such as chaperones or extrinsic stabilising compounds, "salvage" nascent polypeptide chains, and what are the structural characteristics of the intermediates is under investigation.

ii Regarding the physical and chemical limits of growth, it is trivial that, given the canonical repertoire of amino acids and the peptide bond, hydrolysis and chemical modification define the pH limits, as well as the upper limit of temperature. At neutral pH, hydrothermal degradation of all natural amino acids except gly, ala, val, leu, ile and phe becomes significant at T≥140°C; at pH 2, the limiting temperature is ≈ 125°C (Bernhardt *et al.* 1984). Since the hydrophobic hydration also seems to vanish at around

120°C - 140°C, both the stability of the building blocks and their interactions are supposed to determine the upper limit of thermophilism. The ultimate requirement of active life has been assumed to be the presence of water in its liquid state. As shown by the vain attempts to verify the occurrence of bacteria in volcanic areas in the deep sea (Bernhardt *et al.* 1984; White 1984; Trent *et al.* 1984), this assumption must be understood within reasonable limits; "black smoker bacteria" belong to Jules Verne's repertoire. Obviously, the limits of halophily are mainly defined by the "compatibility" of the salts involved; organisms exposed to incompatible salts adjust isosmotic conditions by ion-specific pumps (exchanging, e.g. Na^+ against K^+), or they synthesise metabolites, this way responding to climate-dependent stress variations (Franks and Mathias 1982; Eisenberg and Wachtel 1987; Jaenicke 1991b; Schmid 1991).

Proteins from Extremophiles

Any sound analysis correlating protein structure and stability has to be based on known three-dimensional structures and a sufficiently large set of protein sequences. So far, such data are only available for proteins from thermophiles. Data for other "extremophilic proteins", especially from acido/alkalophilic and barophilic organisms, are scarce, if available at all. In the case of proteins from halophiles, certain characteristics are well-established (Eisenberg and Wachtel 1987; Eisenberg *et al.* in press), however no high-resolution data have been put forward so far. In order to maintain structural integrity at low water activity, proteins have to compete for their hydration with the highly concentrated electrolyte solution. To accomplish this, halophilic proteins favor strongly hydrated amino acids such as glutamic acid and arginine; at the same time, they show a preference for polar instead of non-polar residues. As in most mechanisms of stress response, the central issue in adaptation is the conservation of flexibility which, at high salt, may be accomplished by a decrease in hydrophobicity. How halophilic proteins fulfil this requirement in terms of three-dimensional structure, is still unresolved. Models assuming a hydrophobic core with hydrophilic loops extending into the solvent contradict homology modelling approaches which, in the case of dihydrofolate reductase, show that the sequence of the halophilic protein fits perfectly into the overall structure of its non-halophilic counterparts (Böhm 1992). The clarification of the contradicting models has to await the result of the high-resolution X-ray analysis.

The specific interest in thermal stability of proteins has its origin in the wide range of questions related to evolutionary adaptation, thermodynamics and biotechnological applications. In this case, the amount of data is immense. However, statistical analyses of both amino-acid compositions and sequences have still been unsuccessful in establishing either a general mechanism of thermophily, or "traffic rules" for the rational design of proteins with enhanced thermal stability. Obviously the structural basis of thermophilic behavior is individual in each case, as one might expect from the previous statement that the marginal $\Delta\Delta G_{stab}$ increment that causes improved stability may be provided by a large number of slightly altered spatial arrangements of the polypeptide chain. Local effects seem to accumulate so that, even in cases where highly homologous proteins are compared, improvements may refer to astronomical numbers of combinations of interactions (Jaenicke 1991c).

What is well-established (and describes the effect more than the cause), is the observation that high intrinsic stability of thermophilic proteins, monitored at room temperature, corresponds to anomalously low conformational flexibility. This is clearly detectable from low exchange rates of amide protons, increased resistance against proteolysis and chaotropic agents, NMR data, electron densities and calculations based on normalised B-values (Wagner and Wüthrich 1979; Frauenfelder *et al.* 1979; Wrba *et al.* 1990a,b). At high temperature, their stability becomes marginal so that homologous enzymes under their respective physiological conditions seem to occupy "corresponding states" characterised by similar flexibility (and catalytic activity).

The above mentioned parabolic temperature profiles of the free energy of stabilisation for thermophilic proteins seem to be flattened rather than shifted. For the few cases where comparative thermodynamic data have been determined, ΔG_{stab} has been reported to be only marginally temperature-dependent, whereas in the case of mesophiles a significant temperature dependence is observed. Generally, enthalpy-entropy compensation is found to operate (Lumry and Rajender 1970). As a consequence of the flattening effect, (hyper-)thermophilic proteins do not exhibit cold denaturation as a common feature (Nojima *et al.* 1978; Jaenicke 1992; Schultes and Jaenicke 1992).

Available structural data show that thermostable proteins are strikingly similar to their mesophilic counterparts as far as their basic topology, activity and enzymatic mechanism are concerned. Identical catalytic mechanisms suggest essentially complete conservation of the amino-acid residues constituting the active site(s). The free energies of ligand binding for coenzymes and substrates do not require thermal adaptation, since, again, the temperature dependences of ΔH and $T\Delta S$ compensate each other. For a detailed discussion of selected examples of (hyper-)thermophilic proteins see Jaenicke 1991b,1992. In summarising the data, among the general mechanisms of protein stabilisation, i.e. amino-acid substitution, ligand binding, crosslinking and association, the first two strategies are most frequently found. In general, stabilising effects have been found to be additive. Extrinsic effects or effectors may enhance or replace intrinsic stabilisation.

Protein Folding and Association

Experiments to Mimic Folding and Association *in Vivo*

Denaturation has not only been a means to monitor protein stability but also the most powerful approach to mimic protein folding. 60 years experience, since Anson performed his first experiments, has shown that proteins may be renatured ("unboiled") by restoring native conditions (Anson 1945; Ghélis and Yon 1982; Jaenicke 1987a). The corresponding folding reactions basically refer to the complete polypeptide chain, whereas *in vivo* the nascent chain may be assumed to fold cotranslationally, i.e. from the N- to the C-terminal end . The fact that both the native protein and the product of renaturation are indistinguishable suggests that the *in vitro* approach mimics the *in vivo* reaction (Jaenicke 1987a). However, in this comparison, factors such as local concentrations within specific cell compartments, cytosolic solvent conditions, cotranslational and posttranslational modification, transcriptional or translational control, kinetic competition of folding and association, have been neglected.

Except for the effects of protein concentration, viscosity and specific ligands, hardly any attempts have been made to investigate possible influences of cytosolic solvent conditions on the folding process. With regard to the above cell biological implications, neither directionality (due to cotranslational "folding by parts"), nor discontinuities in chain elongation can play a significant role (Jaenicke 1988).

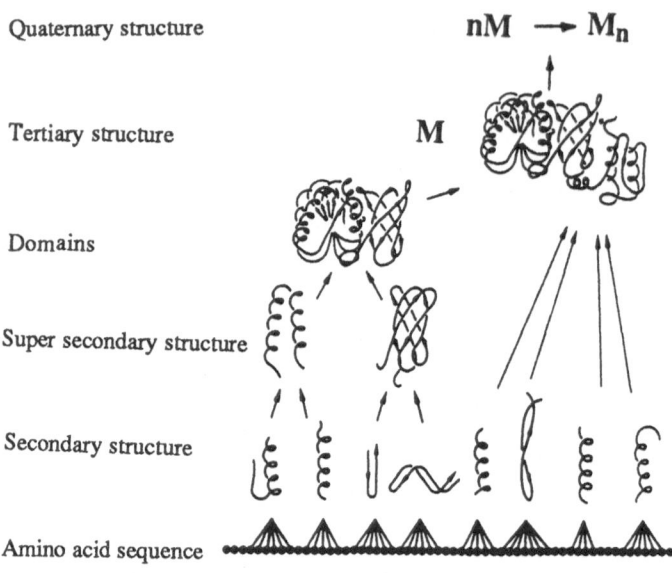

Quaternary structure

Tertiary structure

Domains

Super secondary structure

Secondary structure

Amino acid sequence

$nM \longrightarrow M_n$

M

Fig. 1. 2. Hierarchy of protein structure and protein folding.

Glycosylation has been shown to have no significant effect on protein folding: in the case of ribonuclease and invertase, neither the structure nor the kinetic mechanism of folding exhibit significant differences when the glycosylated, core-glycosylated and non-glycosylated proteins are compared (Krebs *et al.* 1983; Kern *et al.* in press). What is affected is the stability and the tendency to aggregate: the carbohydrate moiety causes increased solubility so that upon secretion glycosylation keeps the nascent polypeptide chain from misassembly or aggregation (Kern *et al.* in press).

Hierarchical Condensation

Considering the hierarchy of protein structure (Fig. 1.2), the kinetic mechanism of protein self-organisation is not a strictly sequential reaction proceeding from local secondary and supersecondary structural elements to specific tertiary contacts defining the native state. The reason is that the tertiary contacts themselves select the correct local structures and stabilise them, such that local structures with low stability and high folding and unfolding rates are in rapid equilibrium early in folding. What seems clearly established in the "hierarchical condensation" or "framework model" of protein folding is the observation that structure formation starts from next-neighbor interactions which, in the time range of milliseconds, generate stretches of secondary structure. Subsequently, non-local interactions lead to metastable supersecondary structures and subdomains which, still within fractions of a second, collapse into the native-like "molten globule state". "Shuffling" to form the native tertiary structure occurs by a limited number of pathways, with the rate-limiting steps as late events. If, at this level, there is still significant hydrophobic surface area exposed to the aqueous solvent, association will end up in the geometrically and stoichiometrically unique quaternary structure. Since various aspects of this "consensus pathway" for sequential protein folding have recently been covered in a number of reviews (Creighton 1978; Ghélis and Yon 1982; Goldberg 1985; Wodak *et al.* 1987; Jaenicke 1987a, 1991a,d; Baldwin 1989; Kim and Baldwin 1990; Christensen and Pain 1990), I shall restrict myself to selected questions of potential technical relevance. They refer to:

i   the biological significance of *in vitro* folding studies;
ii   the kinetics and yield of protein folding;
iii   crosslinking and immobilisation; and
iv   the code of protein folding.

Biological Significance

In summarising the evidence that *in vitro* folding yields "the protein in its native state", and that the pathway of *in vitro* folding simulates the pathway in the cell, a first argument comes from the identity of the global properties of the initial cellular compounds and their reconstitution products (Jaenicke 1987a). In many instances, the full reversibility of the $U \rightleftharpoons N$ equilibrium transition from the unfolded (U) to the renatured (native) state (N) proved the folding transition to be a two-state process. Under certain conditions, stable intermediates could be deduced from biphasic or multiphasic kinetics attributable to specific steps in the previously mentioned structural hierarchy (Kim and Baldwin 1982, 1990). In this context, a most powerful approach has been the crosslinking reaction of intrachain cystine bridges during the folding of native disulfide-bonded protein molecules. Using this technique, the folding pathway of small polypeptide chains, as well as cotranslational domain folding have been accessible to a detailed kinetic analysis (Creighton 1978; Bergman and Kuehl 1979; Freedman 1984).

Because of the unique structure-function relationship of proteins, the biological (enzymatic) activity may be used as a sensitive criterion for the correctness of the "renativated" structure. That reconstituted enzymes frequently exhibit activities exceeding the starting material is easily explained by the fact that the denaturation-renaturation cycle just adds another purification step. The alternative explanation (which has also been applied to incomplete reactivation) that a given protein molecule forms multiple well-defined thermodynamically stable states with varying functional properties is not supported by sound experiments.

In connection with the domain structure of increasingly larger polypetide chains, consecutive "folding-by-parts" confirms the hypothesis that domains may unfold and fold as independent entities. For example, in the case of γ-crystallin, the N-terminal domain unfolds at higher denaturant concentrations than the C-terminal domain; upon refolding under quasi-physiological conditions, the more stable N-terminal domain refolds fast, giving rise to the halfstructured intermediate which, in the subsequent slower step, regains its correct native structure (Rudolph *et al.* 1990; Sharma *et al.* 1990). For the analogous *in vivo* folding process, immunoglobulin translation in mouse myeloma cells has been used, with the result that intra-domain crosslinking occurs sequentially, one domain after the other, leaving space for mutual stabilisation by domain merging after translation has come to its end (Bergman and Kuehl 1979).

The specificity of domain pairing and subunit recognition, even in the presence of excess heterologous proteins, has been shown for a number of systems (Richards and Vithayathil 1959; Goldberg and Zetina 1980; Wetlaufer 1981; Jaenicke *et al.* 1981; Gerl *et al.* 1985; Jaenicke 1987a; Opitz *et al.* 1987; Rudolph *et al.* 1990). Experiments

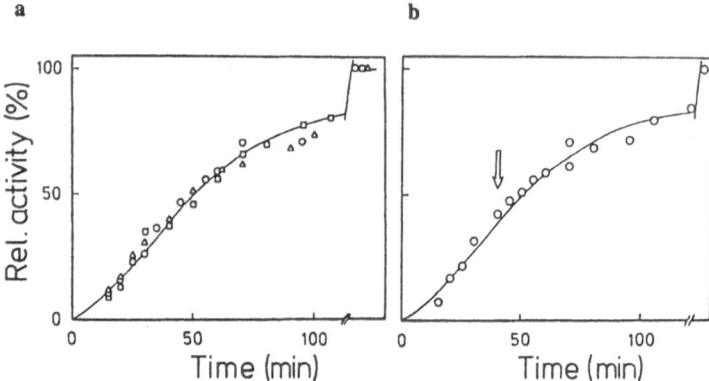

Fig. 1.3. Specificity of selfassembly of (dimeric) malate and lactate dehydrogenases (Gerl *et al.* 1985). Joint (**a**) and separate (**b**) reactivation of Limulus D-LDH and pig m-MDH in 0.1 M phosphate pH 7.5, 20°C, after simultaneous denaturation in 6.4M guanidinium chloride. (**a**) Reactivation of D-LDH (4 μg/ml) in the presence of 2 μg/ml (△), 0.52 μg/ml (□), 0.2 μg/ml (○) m-MDH. (**b**) Reactivation of D-LDH (4 μg/ml) in the absence of m-MDH; after 40 min (arrow), denatured m-MDH (c = 2 μg/ml) was added. No subunit exchange after long-term incubation. Since there is no mutual effect on the yield and kinetics, chimeric intermediates can also be excluded. For similar experiments with pig-heart LDH-H_4, cf. (Jaenicke *et al.* 1981).

of this type simulate the situation in the cytoplasm where numerous nascent polypeptide chains fold synchronously, obviously without forming incorrect (chimeric) assembly structures (Fig. 1.3). Of course, one has to keep in mind that there are many examples, e.g. in the case of multiple forms of enzymes (isoenzymes of LDH etc), where hybrids with specific metabolic properties are formed in the cell. On the other hand, wrong domain pairing cannot be excluded; it may be responsible for incomplete reconstitution of large monomeric proteins (Jaenicke 1987a).

At the highest level of the structural organisation, intrinsic structure determination is restricted to the more simple regular multimeric systems. For example, in the case of the two-dimensional assembly of bacterial surface layers (see Pum *et al.* chapter 10 this volume), *in vitro* experiments under appropriate conditions yield quasinative superstructures, without the requirement of any additional components (Jaenicke *et al.* 1985; Hecht *et al.* 1986). The same holds for simple viruses (Jaenicke 1987a). Complex assembly systems such as ribosomes or phages need sophisticated multistep procedures in order to mimic the natural sequential process, which in this case is determined by the genome organisation, as well as helper proteins (Nierhaus 1990).

The complex assembly systems clearly show that cellbiological implications may be significant in protein folding and association. To mention only a few, folding *in vivo* may be affected by: genetically determined morphopoiesis, vectorial cotranscriptional effects, codon usage, non-linear translation rate, cytoplasmic factors such as ions, cofactors, chaperones, catalysis of slow steps (e.g. by protein disulfide isomerase (PDI) or peptidyl prolyl *cis-trans* isomerase (PPI)), and chemical modification of the polypeptide chain by glycosylation, proteolytic trimming, splicing, etc. Some of these potential effects may be of importance; clear proof has been presented for morphopoietic factors, cotranscriptional effects, PDI catalysis and chemical modification (Freedman 1984,1989; Jaenicke 1987a, 1991a; Nierhaus 1990; Fischer and Schmid 1990; Schmid 1991; Schmid *et al.* in press; Kern *et al.* in press).

Folding Kinetics and Catalysis of Folding

The synthesis of a polypeptide chain of average length takes of the order of 1 min. The rate of *in vitro* folding varies in a wide range. In the case of small proteins without disulfide bridges reactivation after complete denaturation ranges from < 1 s (Staphylococcal nuclease and aldolase) to 10 s (myoglobin). Thus, in these cases translation is rate-determining, and folding and synthesis may be assumed to occur in a synchronised fashion. For small crosslinked proteins, such as (oxidised) ribonuclease, the same holds true; reoxidation, however, takes >10 min. In this case, the common *in vitro* approach differs drastically from the *in vivo* process, which shows parallel folding and crosslinking: translation is followed by:

i   (PDI catalysed) crosslinking;
ii   translocation through the membrane (cleavage by signal-peptidase); and
iii secretion.

There is no detailed information regarding the folding pathway *in vivo*. *In vitro* kinetics have clearly proven that the characteristic features of the chain fold are formed early on the folding pathway (Staley and Kim 1990); proline cis-trans isomerisation is the last (rate-limiting) step (Kim and Baldwin 1982;Fischer and Schmid 1990; Kiefhaber *et al.* 1990; Schmid *et al.* in press).

Large polypeptides without cystine bridges may still exhibit fast folding, whereas large crosslinked proteins, due to slow disulfide exchange reactions, generally show exceedingly slow kinetics (Buchner and Rudolph 1991). In both cases, *in vivo* and *in vitro* studies give evidence that incompletely folded intermediates "present" correct epitopes to conformation-specific antibodies. *In vitro*, a great variety of specific (intrinsic and extrinsic) labels have been applied in order to resolve the multiphasic reaction sequence, but still, there is no complete set of data which would allow a detailed description of the single steps along the overall folding pathway of any protein. It is trivial that this statement holds even more for the process *in vivo*. As in the case of the additivity of increments of stability, a promising approach to reach this goal are investigations making use of short peptides (Wright *et al.* 1988; Creighton 1988; Vita *et al.* 1989; Baldwin 1990; Kim and Baldwin 1990; Staley and Kim 1990).

The kinetics of protein folding depend on a great variety of parameters. They have commonly been chosen such that folding intermediates are sufficiently populated to be accurately detectable. This means that potential effects of *in vivo* conditions have been widely ignored so far. Apart from the common solvent parameters and the local concentration of the nascent polypeptide, they refer to specific ligands, catalysts for rate-limiting steps and chaperones that might "assist" in the folding process.

Varying the temperature may alter the folding pattern qualitatively: At low temperature, proline isomerisation (due to its high activation energy) is expected to be rate-determining, whereas under thermophilic conditions its influence may vanish; in contrast, endothermic processes such as certain assembly reactions only take place at

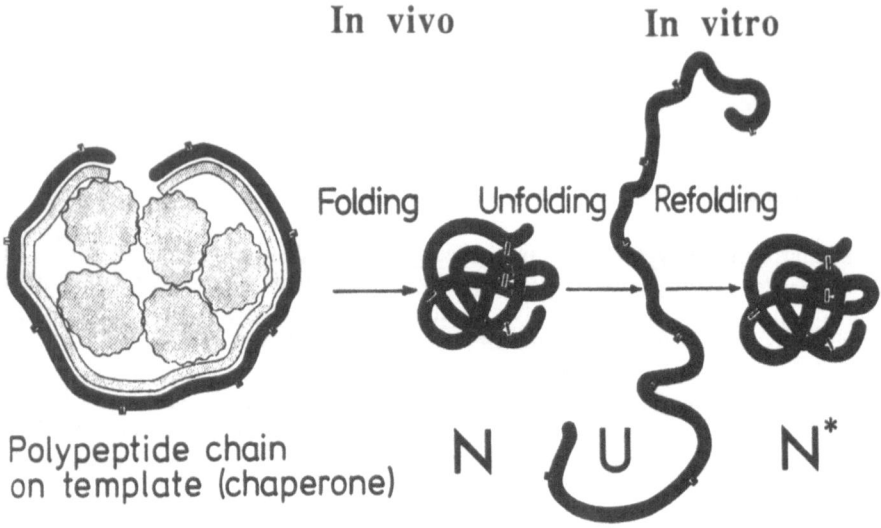

Fig. 1.4. Folding *in vivo* versus folding *in vitro* according to Anfinsen (Epstein *et al.* 1963). In the case of ribonuclease, unfolding and refolding include reduction (by mercaptoethanol at high urea concentration) and reoxidation (e.g. by oxygen).

high temperature so that at 0°C stable intermediates become detectable which might not be populated in the cell (Rudolph *et al.* 1978; Jaenicke *et al.* 1980; Nierhaus 1990).

Ligands may affect folding by stabilising intermediates and/or the final state, this way shifting the $U \rightleftharpoons N$ equilibrium to the right hand side. The effect of Zn^{2+} or the coenzyme, NAD^+, on the folding and association of liver alcohol dehydrogenase and glyceraldehyde-3-phosphate dehydrogenase may serve as examples (Rudolph *et al.* 1978; Jaenicke *et al.* 1980). In many cases, coenzymes lack an effect because their tight binding requires the folding process to have more or less reached the native state. This holds especially in cases where the active site(s) are shared between domains or subunits.

Folding catalysts were postulated and discovered early in the folding game (Goldberger *et al.* 1963; Venetianer and Straub 1963) (Fig. 1.4), but they were ignored for more than two decades. Their importance has only recently become clear, mainly because of two observations: First, experiments with PDI-depleted dog microsomes have shown that protein disulfide isomerase is essential for protein folding in the endoplasmic reticulum (ER) (Freedman 1984, 1989); second, peptidyl prolyl cis-trans isomerase (which has been demonstrated to catalyse proline isomerisation as a late event on the folding path (Kiefhaber *et al.* 1990)), is identical with cyclophilin, the binding protein for the immunorepressant cyclosporin A (Fischer *et al.* 1989). The possible role of proline isomerisation in the immuneresponse is unclear; hypothetical mechanisms have been summarised in (Fischer and Schmid 1990; Schmid 1991; Schmid *et al.* in press).

In this connection molecular chaperones need consideration. Their definition as "mediators of correct folding which transiently interact with nascent polypeptide chains without being components of the final functional state" (Fischer and Schmid 1990; Ellis and van der Vies 1991), puts them close to catalysts although they seem to be

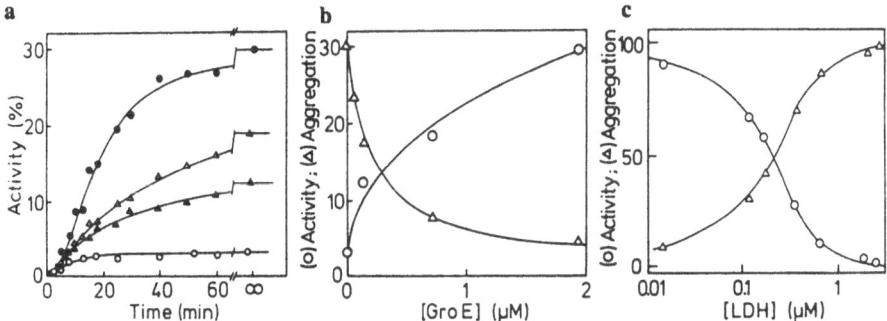

Fig. 1.5. Effect of enzyme concentration and GroE on the reconstitution of citrate synthase and lactate dehydrogenase (Zettlmeißl *et al.* 1979; Buchner *et al.* 1991). (a) Reactivation of 0.3 μM CS in the presence of 1.9 (●), 0.7 (▲), 0.14 (▲) and 0 (o) μM GroE in 0.1 M Tris pH 8, 10 mM KCl, 2 mM ATP, 25°C after denaturation in 6 M guanidinium chloride. (b) Effect of GroE concentration on the final yield of reactivation (o) and aggregation (▲) (monitored by light scattering); conditions as in (a). (c) Kinetic competition of reconstitution (o) and aggregation (▲) for pig muscle $LDH-M_4$ after denaturation in 1 M glycine/H_3PO_4 and renaturation at neutral pH, 10°C (Zettlmeißl *et al.* 1979).

involved in the inhibition of premature folding and association rather than conventional catalysis. The way how intermolecular protein-protein interactions promote correct intramolecular interactions is difficult to visualise, and presently, no mechanistic details regarding the mechanism of chaperones are known. The following examples may serve to illustrate certain principles:

i   The recognition signal in the case of the signal recognition particle is a repeating pattern of methionine residues that are assumed to form a "bristle" accomodating a wide variety of hydrophobic sequences with low sequence specificity.

ii The GroE complex from *E. coli* facilitates the *in vitro* reconstitution of a number of proteins, in most cases requiring ATP and K^+ for the release of the refolding polypeptide chain. Using light scattering and reactivation to monitor the reaction, it becomes clear that GroE inhibits aggregation as the competitive side reaction of correct folding (see below) (Fig. 1.5). Once formed, aggregates are not dissolved by the chaperone (Buchner *et al.* 1991).

iii The only high-resolution crystal structure of a chaperone presently available (PapD) shows a hydrophobic crevice between its two domains as the best candidate for intermediary binding of the nascent polypeptide chain. The fact that bovine serum albumin has been found to be almost as efficient as true chaperones in improving the reconstitution yield fits the previous result since BSA is known to exhibit hydrophobic binding sites on its surface as well.

iv There is some evidence that the size and structure of the target protein dictates the release from the chaperone: obviously, proteins incapable of correct folding are trapped and finally subjected to degradation. In this context, chaperones seem to be part of a larger complex connected with general "editing functions".

Yield

As has been mentioned, hierarchical condensation goes through intermediate states where hydrophobic residues are still exposed to the solvent. This implies that they can easily interact intermolecularly to form aggregates instead of the native tertiary structure (Mitraki and King 1989). As any isomerisation, protein folding is a first-order reaction, whereas aggregation is of higher than second-order (Zettlmeißl *et al.* 1979). Thus, beyond a limiting protein concentration, aggregation will compete with correct folding, explaining two observations: first, why commonly reconstitution can only be accomplished at low protein concentration, and, second, why overexpression of recombinant proteins in most cases yields inclusion bodies. The two phenomena clearly show that the competition of aggregation and folding is important both *in vitro* and *in vivo*. They can be quantitatively described by the superposition of the rate-limiting first-order folding reaction and diffusion-controlled second-order aggregation, according to

$$A \xleftarrow{\quad k_2 \quad} U \xrightarrow{\quad k_1 \quad} N$$

Correspondingly, reactivation yields can be computed as a function of the initial concentration of the denatured protein (Kiefhaber *et al.* 1991).

As indicated, both aggregates and inclusion bodies derive from partially folded intermediates rather than unfolded or mature native protein. Apart from the recombinant gene product, inclusion bodies may contain proteins from the host, as well as nucleic acids. For a great variety of recombinant proteins, including molecules as complex as tissue plasminogen activator, the solubilisation and the overall down-stream processing has been worked out with unpredicted success (Rudolph 1990; Buchner and Rudolph 1991; Kiefhaber *et al.* 1991).

Generally, protein biosynthesis and *in vivo* structure formation is assumed to yield 100% native protein without requiring extrinsic factors or components. The following four observations show that this assumption may not always be true:

i   Protein structure not only controls function but also compartmentation and turnover. During secretion, misfolded, misassembled and unassembled polypeptides are retained in the ER and specifically degraded. This "quality control" by the ER leads to the apparent yield of 100% (Hurtley and Helenius 1989).

ii   As shown for the phage P22 tailspike protein, even under optimum growth conditions of the bacterium, the yield of *in vivo* folding is <50%; under unbalanced physiological conditions, wrong conformers are produced which are continuously removed by proteolysis.

iii Chaperones are involved in folding-association, regulating the yield in an ATP-dependent way.

iv   Overexpression of recombinant protein leads in most cases to wrong aggregation and deposition of the nascent protein in inclusion bodies. In summarising these observations and the underlying mechanisms, it is obvious that both folding *in vivo* and *in vitro* are frequently incomplete, most probably due to the same competing mechanism.

Oligomers : Association

The early stages during the (re-)folding of oligomeric proteins are expected to be identical with those involved in the self-organisation of single-chain proteins. The final product is the native-like "structured monomer" which in a subsequent association reaction is transformed into the native quaternary structure so that the overall process

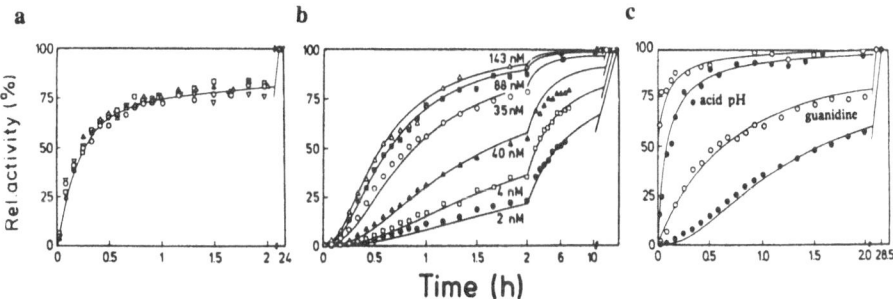

Time (h)

Fig. 1.6. Kinetic mechanism of the reconstitution of malate and lactate dehydrogenase. (a) Cytoplasmic MDH in phosphate buffer pH 7.6 after denaturation in 6 M guanidinium chloride, 20°C, at 1-13 μg/ml. (b) Mitochondrial MDH under conditions as in (a), at 0.07 (●), 0.14 (□), 0.35 (▲), 1.2 (o), 3.1 (■) and 5.0 μg/ml (Δ). (c) Muscle LDH-M$_4$ in phosphate buffer pH 7.6 after denaturation in 1 M glycine/H$_3$PO$_4$ pH 2 or 6 M guanidinium chloride. ●/o , activity in 0 M / 1.5 M ammonium sulfate. Profiles after acid denaturation calculated for dimer-tetramer transition (only k$_2$) with 0/50% subunit activity, those after guanidine denaturation for uni-bimolecular mechanism (k$_1$,k$_2$) with 0% activity (Jaenicke 1987a).

may be written as a sequence of (unimolecular) folding reactions and bimolecular association steps according to

$$n\mathcal{M} \xrightarrow{k_1} nM \xrightarrow{k_2} M_n$$
$$\downarrow k_{>2}$$
$$A$$

where n is the number of subunits, and $\mathcal{M}$ and M the fully denatured and "structured" monomer, M$_n$ the native assembly, and A the aggregated state. The latter is included since, in the given mechanism, kinetic competition between folding and association is preconceived simply by the fact that, at sufficiently high concentration, bimolecular association will outrun folding so that aggregation becomes the dominant reaction (Fig. 1.5c). Whether folding (k$_1$) or association (k$_2$) are rate-determining in the sequential uni-bimolecular reaction depends on the protein, as well as the state of denaturation (fully denatured or partially denatured A-state, or structured monomer). Taking dimeric malate dehydrogenase (MDH) as an example, Fig. 1.6a and 1.6b show that the two isoenzymes represent the extremes: Mitochondrial MDH exhibits sigmoidal and concentration dependent reactivation kinetics (characteristic for the sequential folding-association mechanism), whereas the cytoplasmic enzyme obeys first-order kinetics (indicating rate-determining folding and "shuffling" at the monomer and dimer level and diffusion-controlled association). As one would predict for the "hierarchical condensation", the formation of the native secondary structure is formed before domains or the complete tertiary and quaternary structure condense to minimise the solvent accessible surface area. Subunit assembly requires "structured monomers"; however, quaternary structure formation does not necessarily follow a sequential uni-bimolecular kinetic mechanism. This conclusion becomes even more obvious in the case of lactate dehydrogenase, where the mechanism of reconstitution is found to be strongly affected by the denaturation conditions. Renaturation after acid denaturation starts from "structured monomers" and follows simple second-order kinetics, whereas renaturation after complete randomisation in 6 M guanidinium chloride is again quantitatively described by the sequential folding-association mechanism (Fig. 1.6c). The kinetic analysis is confirmed by crosslinking experiments which allow a snap-shot analysis of assembly intermediates on the kinetic pathway of reconstitution. Applying HPLC, light scattering and ultracentrifugation, the assembly mechanism of particles up to the size of ferritin has been elucidated. In the latter case, the assembly kinetics follow an extended form of the above sequential uni-bimolecular kinetic scheme, with monomers, dimers, trimers and hexamers as intermediates (Jaenicke 1987a; Gerl et al. 1988).

It is plausible to assume similar mechanisms *in vivo* where only few data have been collected so far. The best-understood example is the tailspike protein from the Salmonella phage P22, where ts- and repressor mutants have allowed to block the reaction, or let it proceed. The genetically determined *in vivo* assembly scheme is fully confirmed by *in vitro* experiments (Seckler and Jaenicke, in press).

Reconstitution of Crosslinked and Immobilised Enzymes

As has been shown, wrong intermolecular interactions are the major cause of side reactions interfering with the complete reconstitution of proteins. On the other hand, covalent crosslinks (normally absent in cytoplasmic proteins, especially as intersubunit linkages) may contribute significantly to protein stability (see above). In connection with the stabilising effect of covalent crosslinks on the quaternary structure of proteins, and the potential of matrix-bound enzymes to undergo reversible denaturation, two approaches have been used:

i   Lactate dehydrogenase was crosslinked by glutardialdehyde preserving its native quaternary structure and at least part of its catalytic activity;
ii   monomeric glucoamylase, fixed to porous and non-porous solid matrices, was investigated regarding its stability, specific activity and denaturation-renaturation properties.

Chemically crosslinked lactate dehydrogenase. Fixing the native tetrameric quaternary structure of lactate dehydrogenase by glutardialdehyde may be optimised such that the product retains 60% of the specific activity of the native enzyme. The stability of the covalently linked 140 kDa particle (with ≈4 intersubunit linkages) is enhanced, whereas reconstitution after preceding denaturation is essentially blocked due to incorrect near-neighbor interactions between the crosslinked subunits. After partial (acid) denaturation, 30% reactivation occurs in a fast first-order reaction, as expected for a single chain molecule with significant residual structure (Gottschalk and Jaenicke 1987). Obviously, there is no simple way to regenerate covalently crosslinked oligomeric proteins, applied, e.g. in enzyme reactors; the well-established denaturation-renaturation procedures (Jaenicke and Rudolph 1988) cannot be used.

Matrix-bound glucoamylase. Glucoamylase II, immobilised to Eupergit C and C1Z as porous and non-porous matrices, shows enzymatic properties indistinguishable from those of the free enzyme, except for reduced specific activity due to non-productive fixation of the enzyme or "crowding" effects at the active site. The denaturation equilibrium transition obeys the two-state mechanism, whereas renaturation is a complex reaction dominated by "entangling" of the unfolded polypeptide chain with the matrix. As in the previous case, the reactivation of the immobilised enzyme shows no concentration dependence (Gottschalk and Jaenicke 1991).

Considering the glucoamylase data as a model for the folding of the nascent polypeptide chain coming off the ribosome, or trapped by a chaperone, it becomes clear that "matrices" assisting protein folding in the cell must show a certain degree of specificity. An indifferent matrix and statistical fixation of the polypeptide chain are inadequate to model chaperone action. It seems worth trying to mimic the vectorial process of protein folding by a topologically specific mode of immobilisation, comparing, e.g. protein constructs with C- or N-terminal polyhistidine "tails", fixed to an ion-exchange column.

Code of Protein Folding

There is a two-step information transfer in protein translation:

1   from the genetic (mRNA) level to the amino-acid sequence; and
2   from the one-dimensional to the three-dimensional structure.

The mechanism at the first level is established; the code to perform the second step is still unknown, for a couple of reasons:

i   the energy minima occupied by common proteins are shallow so that energy minimisation deals with the problem of small differences of big numbers;
ii   many independent variables determine the conformation of a complex polyatomic molecule;
iii   next-neighbour and through-space interactions determine the potential energy surface;
iv   the weak interactions involved in protein stability are not fully understood, especially concerning hydrophobicity, hydration, dielectric properties, electrostatics;
v   structure predictions based on crystallisable proteins are biased;
vi   the code is highly degenerate;
vii   co- and posttranslational processing and conjugation of the polypeptide chain may be of importance;
viii cell-biological implications, not inherent in the amino-acid sequence may not only determine the pathway but also the product of folding.

In spite of these intrinsic problems, there have been numerous attempts to forecast the three-dimensional structure of proteins or their folding mode, based merely on the amino-acid sequence (Anfinsen and Scheraga 1975; Blundell et al. 1987; Jaenicke 1988; Fasman 1989; Gierasch and King 1990; Sali et al. 1990; Overington et al. 1990): Search programs for sequence homologies have been applied to correlate given primary structures to a limited number of protein families; statistical analyses of preferences for certain local structural elements in known three-dimensional structures have been used to estimate the secondary structure; topological considerations and docking procedures have been developed to optimise minimum hydrophobic surface area and maximum packing density; energy minimisation and molecular dynamics calculations, as well as quantum-mechanical and statistical mechanical methods have been applied to reduce the number of possible conformations etc. A combination of all available methods in terms of "knowledge-based computer-aided structure predictions" has provided structure predictions with root-mean-square deviations < 3 Å (Blundell *et al.* 1987; Sali *et al.* 1990; Overington *et al.* 1990). It is obvious that forecasts with regard to the correlation of structure, function, mechanism and specificity require still higher precision. More high-resolution X-ray data and their detailed statistical evaluation, as well as methods derived, e.g. from polymer statistics, pattern recognition techniques or neuronal networks, may provide new concepts to eventually approach a solution to the protein folding problem.

Acknowledgements

I thank the Fogarty International Center for Advanced Study, National Institutes of Health, Bethesda, MD, for generous support and hospitality. Fruitful discussions with Drs. G. Böhm, K. Kirschner, P. L. Privalov, R. Rudolph, F.X. Schmid and R. Seckler are gratefully acknowledged. Work performed in the author's laboratory was generously supported by the Deutsche Forschungsgemeinschaft and the BAP Program of the European Commission.

References

Anderson DE, Becktel WJ, Dahlquist FW (1990) Biochemistry 29: 2403-2408
Anfinsen CB, Scheraga HA (1975) Adv Prot Chem 29, 205-300
Anson ML (1945) Adv Protein Chem 2: 361-384
Baldwin RL (1989) Trends Biochem Sci. 14: 291-294
Baldwin RL (1990) In: Protein Design and the Development of New Therapeutics and Vaccines (Hock JB and Poste G Eds) Plenum, New York 49-57
Baldwin RL, Eisenberg D (1987) In: Protein Engineering (Oxender DE and Fox CF Eds) Liss, New York 127-148
Barlow DJ, Thornton JM (1983) J Mol Biol 168: 867-885
Bergman LW, Kuehl WM (1979) J Supramol Struct 11: 9-24
Bernhardt G, Lüdemann H-D, Jaenicke R, König H, Stetter KO (1984) Naturwissenschaften 71: 583-586
Blundell TL, Sibanda BL, Sternberg MJE, Thornton JM (1987) Nature 326: 347-352
Böhm G (1992) Dissertation, Universität Regensburg
Buchner J, Renner M, Lilie H, Hinz H-J, Jaenicke R, Kiefhaber T, Rudolph R (1991) Biochemistry 30: 6922-6929
Buchner J, Rudolph R (1991) Bio/Technology 9: 157-162
Buchner J, Schmidt M, Fuchs M, Jaenicke R, Rudolph R, Schmid FX, Kiefhaber T (1991) Biochemistry 30: 1586-1591
Christensen H, Pain RH (1990) Eur J Biophys 19: 221-229
Creighton TE (1978) Progr Biophys Mol Biol 33: 231-297
Creighton TE (1988a) BioEssays 8: 57-63
Creighton TE (1988b) Biophys Chem 31: 155-162
Creighton TE (1991) Curr Opin Struct Biol 1: 5-16
Dill KA (1990) Biochemistry 29: 7133-7155
Eisenberg H, Mevarech M, Zacchai G Adv Protein Chem, in press
Eisenberg H, Wachtel EJ (1987) Annu Rev Biophys Biophys Chem 16: 69-92
Ellis RJ, van der Vies SM (1991) Annu Rev Biochem 60: 321-347
Epstein CJ, Goldberger RF, Anfinsen CB (1963) Cold Spring Harb Symp Quant Biol 28: 439-449
Fasman GD (1989)(Ed) Prediction of Protein Structure and the Principles of Protein Conformation, Plenum, New York, London p 798
Fischer G, Schmid FX (1990) Biochemistry 29: 2205-2212
Fischer G, Wittmann-Liebold B, Lang K, Kiefhaber T, Schmid FX (1989) Nature 337: 476-478
Franks F, Mathias S (Eds)(1982) Biophysics of Water, Wiley p 400
Frauenfelder H, Petsko GA, Tsernoglou D (1979) Nature 280: 558 563
Freedman RB (1984) Trends Biochem Sci 9: 438-441

Freedman RB(1989) Cell 57: 1069-1072
Gerl M, Jaenicke R, Smith JM, Harrison PM (1988) Biochemistry 27: 4089-4096
Gerl M, Rudolph R, Jaenicke R (1985) Biol Chem Hoppe-Seyler 366: 447-454
Ghélis C, Yon J (1982) Protein Folding, Academic Press, New York, London p 562
Gierasch LM, King J (1990)(Eds) Protein Folding: Deciphering the Second Half of the Genetic Code, AAAS, Washington p 334
Goldberg ME (1985) Trends Biochem Sci 10: 388-391
Goldberg ME, Zetina CR (1980) In: Protein Folding (Jaenicke R Ed) Elsevier-North Holland, Amsterdam 469-484
Goldberger RF, Epstein CJ, Anfinsen CB (1963) J Biol Chem 238: 628-635
Goto Y, Takahashi N, Fink AL (1990) Biochemistry 29: 3480-3488
Gottschalk N, Jaenicke R (1987) Biotechnol Appl Biochem 9: 1876-1879
Gottschalk N, Jaenicke R (1991) Biotechnol Appl Biochem 14: 324-335
Hecht K, Wieland F, Jaenicke R (1986) Biol Chem Hoppe-Seyler 367: 33-38
Hensel R, König H (1988) FEMS Microbiol Lett 49: 75-79
Huber R, Kurr M, Jannasch HW, Stetter KO (1989) Nature 342: 833-834
Hurtley SM, Helenius A (1989) Annu Rev Cell Biol 5: 277-307
Jaenicke R (1981) Annu Rev Biophys Bioeng 10: 1-67
Jaenicke R (1987a) Progr Biophys Mol Biol 49: 117-237
Jaenicke R (1987b) In: Current Topics in High-pressure Biology (Jannasch HW, Marquis RE and Zimmerman AM Eds), Academic Press, London, 257-272
Jaenicke R (1988) Colloq Ges Biol Chem Mosbach 39: 16-36
Jaenicke R (1990) Phil Trans R Soc Lond B326: 535-553
Jaenicke R (1991a) Biochemistry 30: 3147-3161
Jaenicke R (1991b) Eur J Biochem 202: 715-728
Jaenicke R (1991c) Ciba Foundation Symp 161: 206-221
Jaenicke R (1991d) In: Applications of Enzyme Biotechnology (Martell AE Ed) Plenum New York, in press
Jaenicke R (1992) In: Biocatalyst Design for Stability and Specificity (Himmel ME , Georgiou G Eds) ACS, New York in press
Jaenicke R, Krebs H, Rudolph R, Woenckhaus C (1980) Proc Natl Acad Sci USA 77: 1966-1969
Jaenicke R, Rudolph R (1988) In: Protein Structure: A Functional Approach (Creighton ET, Ed) IRL Press Oxford 191-223
Jaenicke R, Rudolph R, Heider I (1981) Biochem Int 2: 23-31
Jaenicke R, Welsch R, Sára M, Sleytr UB (1985) Biol Chem Hoppe Seyler 366: 663-667
Kauzmann W (1959) Adv Protein Chem 14: 1-63
Kellenberger E (1984) Helv Phys Acta 57 : 188-201
Kern G, Schülke N, Schmid FX, Jaenicke R. Protein Science, in press
Kiefhaber T, Grunert H-P, Hahn U, Schmid FX (1990) Biochemistry 29: 6475-6480
Kiefhaber T, Rudolph R, Kohler H-H, Buchner J (1991) Bio/Technology 9: 825-829
Kim PS, Baldwin RL (1982) Annu Rev Biochem 51: 459-489
Kim PS, Baldwin RL (1990) Annu Rev Biochem 59: 631-660
Krebs H, Schmid FX, Jaenicke R (1983) J Mol Biol 169: 619-635
Lumry R, Rajender S (1970) Biopolymers 9: 1125-1227
Mitraki A, King J (1989) Bio/Technology 7: 690-697
Nierhaus KH (1990) In: Ribosomes and Protein Synthesis: A Practical Approach (Spedding G, Ed) IRL Press Oxford 161-189
Nojima H, Hon-nami K, Oshima T, Noda H (1978) J Mol Biol 122: 33-42
Opitz U, Rudolph R, Jaenicke R, Ericsson L, Neurath H (1987) Biochemistry 26: 1399-1406
Overington JP, Johnson MS, Sali A,Blundell TL (1990) Proc R Soc Lond B 241: 132-145
Pace CN (1990) Trends Biochem Sci 15: 14-17
Pace CN, Heinemann U, Hahn U, Saenger W (1991) Angew Chem Int Ed Engl 30: 343-360
Pace CN, Laurents DV, Thompson JA (1990) Biochemistry 29: 2564-2572
Perutz MF, Raidt H (1975) Nature 255: 256-259
Pfeil W (1986) In: Thermodynamic Data for Biochemistry and Biotechnology (Hinz H-J Ed) Springer Verlag, Berlin, Heidelberg, New York, 349-376
Privalov PL (1990) CRC Crit Rev Biochem Mol Biol 25: 281-306
Privalov PL, Gill SJ (1989) Pure Appl Chem 61: 1097-1104
Richards FM, Vithayathil PJ (1959) J Biol Chem 234: 1459-1464
Rudolph R (1990) In: Modern Methods in Protein and Nucleic Acid Research (Tschesche H, Ed), de Gruyter, Berlin 149-171

Rudolph R, Gerschitz J, Jaenicke R (1978) Eur J Biochem 87: 601-606
Rudolph R, Siebendritt R, Nesslauer G, Sharma AK, Jaenicke R (1990) Proc Natl Acad Sci USA 87: 4625-4629
Sali A, Overington JP, Johnson MS, Blundell TL (1990) Trends Biochem Sci 15: 235-240
Schade BC, Lüdemann H-D, Jaenicke R (1980) Biochemistry 19: 1121-1126
Schmid FX (1991) Curr Opin Struct Biol 1: 36-41
Schmid FX, Mayr L, Mücke M, Schönbrunner ER, Adv Protein Chem, in press
Schultes V, Jaenicke R, J Biol Chem, in press
Seckler R, Jaenicke R, FASEB J, in press
Sharma AK, Minke-Gogl V, Gohl P, Siebendritt R, Jaenicke R, Rudolph R (1990) Eur J Biochem 194: 603-609
Sharp KA, Honig B (1990) Annu Rev Biophys Biophys Chem 19: 301-332
Shirley BA, Stanssens P, Hahn U, Pace CN, Biochemistry, in press
Staley JP, Kim PS (1990) Nature 344: 685-688
Stigter D, Dill KA (1990) Biochemistry 29: 1262-1271
Teschner W, Rudolph R (1989) Biochem J 260: 583-587
Timasheff SN, Arakawa T (1989) In: Protein Structure: A Practical Approach (Creighton TE Ed) IRL Press Oxford 331-345
Trent JD, Chastain RA, Yayanos AA (1984) Nature 307: 737-740
Venetianer P, Straub FB (1963) Biochim Biophys Acta 67: 166-168
Vita C, Fontana A, Jaenicke R (1989) Eur J Biochem 183: 513-518
Wagner G, Wüthrich K (1979) J Mol Biol 130: 31-37
Weissman JS, Kim PS (1991) Science 253: 1386-1393
Wetlaufer DB (1981) Adv Prot Chem 34: 61-92
Wetzel R, Perry LJ, Mulkerrin MG, Randall LM (1990) In: Protein Design and the Development of New Therapeutics and Vaccines (Hook JB and Poste G Eds), Plenum, New York 79-115
White RH (1984) Nature 310: 430-432
Wodak SJ, de Crombrugghe M, Janin J (1987) Prog Biophys Mol Biol 49: 29-63
Wrba A, Jaenicke R, Huber R, Stetter KO (1990a) Eur J Biochem 188: 195-201
Wrba A, Schweiger A, Schultes V, Jaenicke R, Závodszky P (1990b) Biochemistry 29: 7585-7592
Wright PE, Dyson H-J, Lerner RA (1988) Biochemistry 27: 7167-7175
Zettlmeißl G, Rudolph R, Jaenicke R (1979) Biochemistry 18: 5567-5571

Chapter 2

Immobilisation of Macromolecules and Cells

S. Birnbaum

Introduction

In the following chapter a basic overview of the methods used for immobilisation, the forms of the immobilised preparations and the operating units in which they are employed are presented. A short discussion of the pros and cons of immobilisation then follows with a more detailed description of some of the methods and applications which I and others at this institute have used. For further, in depth, reviews on immobilised macromolecules and cells the reader is referred to some of the recent literature (Birnbaum *et al.* 1983; Mosbach 1987a,b; Mosbach 1988; Taylor 1991; Powell 1991, Birnbaum and Mosbach 1991; Salter and Kell 1991).

The macromolecules which have been immobilised include proteins, nucleic acids, carbohydrates and lipid structures. Proteins are the major class among these macromolecules which have been immobilised, studied and employed. As such, the remainder of this chapter will focus on proteins when discussing immobilisation of macromolecules. The other primary entity of interest for immobilisation are cells. Organelles, an intermediate in complexity between the afore mentioned macromolecules and cells have also been studied but to a significantly lesser degree and therefore are not covered in this chapter.

The use of immobilised proteins and cells is the same as their application when employed in the free form, that is to bind other substances and to catalyse reactions specifically. Thus, an immobilised preparation can be employed for production, purification or analysis of a substance. The impetus to use proteins and/or cells in the immobilised form stems from the advantages obtained by immobilisation in comparison to the free protein or cell as discussed below.

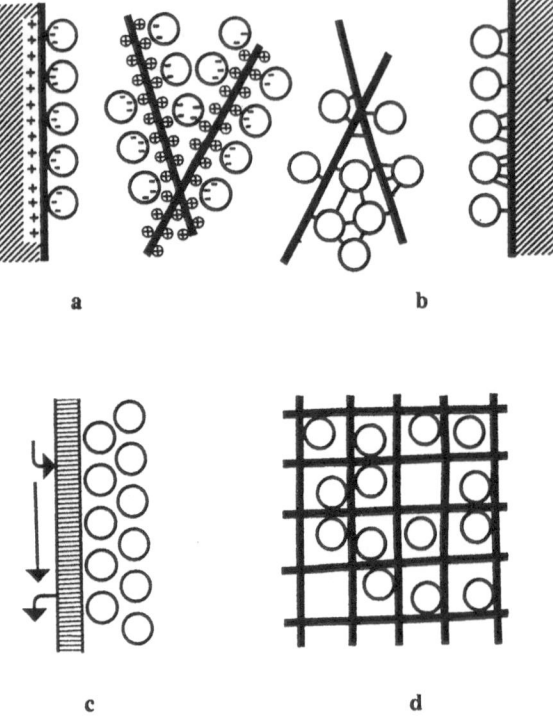

Fig. 2.1a,b,c,d. Methods of immobilisation. **a** Adsorption **b** Covalent coupling **c** Encapsulation **d** Entrapment.

Table 2.1. Representative list of various materials and derivatives there of, used for immobilisation

agar	agarsose	albumin	alginate	alumina
amberlite	brick	carrageenan	celite	cellulose
cellophane	ceramic	chitin	collagen	dextran
epoxy resin	eudragit	fibrin	gelatin	glass
nylon	pectate	polyacrylate	polyester	PEG
polystyrene	polysulphone	polyurethane	PVA	PVC
PVDF	silica	silk	sponge	starch
steel	titanium			

Preparation and Operation

The term immobilisation connotes the physical confinement to a defined spatial region. In general, this has meant that a protein or cell is associated with a support material, either in soluble or insoluble form, which limits its free movement. Various materials of either natural or synthetic origin and composed of either organic or inorganic material have been used for immobilisation. A representative list of the materials used, though not complete is given in Table 2.1. The various methods of immobilisation are depicted in Fig. 2.1. They can be grouped into those based on adsorption, covalent coupling, entrapment or encapsulation. Combinations of various methods may also occur.

Adsorption is primarily achieved through ionic or hydrophobic interactions between the protein or cell and the support. The method is usually very simple, merely mixing the two under the appropriate conditions suffice, and is generally mild. An advantage of the adsorption method of immobilisation is that the support is often regenerable since the binding is reversible. However, the reversibility of binding to the adsorbent also limits this method as the macromolecule or cell may detach from the support, particularly if the optimal conditions during operation are significantly different from those used during adsorption. Some of the more recent methods which have been reported include the use of fluorocarbon supports to which perfluoroalkylated proteins adsorb strongly either through hydrophobic or fluorophilic interactions (Kobos et al. 1989; Stewart et al. 1990).

A particular form of adsorption includes what can be referred to as affinity binding between a protein or cell and its support. In this case, the binding forces usually involves several, often different non-covalent forces including both ionic or hydrophobic interactions but even others such as hydrogen binding and van der Waals interactions. Illustrative examples of the different types of affinity adsorption includes those often found in biological systems where protein-ligand or protein-protein interactions occur, such as enzyme - inhibitor, avidin - biotin, protein A - IgG, or cell - protein interactions. Usually, one of the interacting components has been previously immobilised through covalent coupling methods (see below) to an inert support. Reviews on recent developments in affinity chromatography can be found in the literature (Scouten 1991; Glad and Larsson 1991)

In the case where adsorption is not a sufficiently adequate method of immobilisation, it may be advisable to covalently couple the protein or cell to the support. Several methods for the formation of covalent bonds between the protein or cell and support exist, and can be found in the literature (Scouten 1987). The protein or cell may even be cross-linked using a bifunctional reagent such as glutaraldehyde. Care must be taken at choosing a particular method as the reactive conditions used may damage the functionality of the immobilised entity. One of the more recent methods described includes protein coupling via azlactone derivatives of the copolymer vinyl dimethyl and methylene-bis acrylamide to which, for example, 400 mg of protein A could be coupled per gram of carrier (Coleman et al. 1990).

The third group of immobilisation methods entails the entrapment of the protein or cell within a polymer network. Whereas covalent coupling is most often used for protein immobilisation, entrapment is more usual for cell immobilisation though recent protein entrapment methods in silica are noteworthy (Glad et al. 1985; Braun et al. 1990). Commonly, entrapment is accomplished by mixing the species to be entrapped with an aqueous solution of the monomer or polymer with subsequent polymerisation or gelation. The first example of this method for immobilising cells in polyacrylamide was reported more than 25 years ago (Mosbach and Mosbach 1966).

Physical confinement may also be achieved through the use of membrane encapsulation. Typical examples of this method include the use of preformed membranes either in the form of hollow-fibres or in the form of ultrafiltration devices (Inloes et al. 1983; Baranov et al. 1989). Alternatively, the membrane can also be formed during or after the immobilisation process, for example, in microencapsulation (Sun 1988).

The immobilised preparation can display different forms and physical characteristics. It can, for example, be immobilised in or to either porous or non-porous material. The selection between the use of porous or non-porous materials depends in part on the characteristics which are required for the specific application. Non-porous materials have the advantage that rapid interactions between the mobile and bulk phase occur as mass transfer limitations are minimal. However, the capacity of the immobilised preparation is considerably higher for porous materials. Furthermore, immobilisation in porous particles will protect the protein or cell from adverse conditions in the bulk phase such as physical damage or proteolytic activity. The choice of optimal pore size is a trade off between the maximum capacity of the support and the diffusion or flow characteristics of the immobilised preparation required or desired. As the pore size decreases, the surface area and thus the capacity of the carrier increases. On the other hand, as the pore size decreases, mass transfer limitation within the particle increase and this may severely limit the effectiveness of the immobilised preparation. A general rule of thumb is that a pore size at least on the order of four to five times the maximum dimension of either the immobilised protein or cell should be chosen when they are immobilised via either adsorption or covalent coupling to a preformed porous support, or of the interacting partner. Of course, when the protein or cell is to be entrapped the pore size must be considerably smaller than the immobilised entity yet large enough to allow free passage of any compounds which are to interact with the immobilised protein or cell. Thus, when proteins, which normally have a maximum dimensions of 50 Å - 100 Å, are covalently immobilised in a porous carrier such as silica, a pore size of 300 Å - 500 Å is appropriate. For the immobilisation of mammalian cells, which have a diameter of approximately 10 μm, the desired pore dimension is about 50 μm when macroporous microcarriers are employed (Nilsson et $al.$ 1986, Nilsson et $al.$ 1988a).

The shape of the preparation can either be in the form of particles (usually spheres) or in the form of fibres or membranes (discs or plates). These shapes represent extremes of the three-dimension, fibres the extreme in 1-D, membranes the extreme in 2-D and spheres the extreme in 3-D. The various forms or characteristics of the immobilised preparation have various attributes dependent on the particular application.

Spherical particles with diameters in the range of a few microns up to several millimetres have been used for immobilisation. They can be used in either batch or continuously stirred tank reactor, or they can be employed in column type reactors in either the packed bed or fluidised bed regime. Spheres are preferred to cubes as lower pressure drops are obtained in packed beds and less attrition in stirred tanks. Particle size is again chosen as a compromise between two competing properties. For non-porous particles, the surface area to volume ratio increases as particle diameter decreases thereby increasing capacity. For porous particles a minimal particle size is desirable to minimise diffusion distances. On the other hand, in a packed bed reactor, the pressure drop over the column increases as the particle diameter decreases. The pressures which most proteins can withstand vary, but are in the range of 100 MPa - 600 MPa (Olson et $al.$ 1989). This limits the particle size as well as the flow rate and column length which can be employed when packed beds are used. To alleviate this dilemma Afeyan et $al.$ have recently described "perfusion chromatography" material consisting of a macroporous carrier of 10 μm - 20 μm diameter in which the macroporous "through pores" (6 000 Å - 8 000 Å) are interconnected with smaller pores (ca. 1 000 Å) to increase the surface area of this material (Afeyan et $al.$ 1990). These larger through pores actually allow convective flow directly through the particle.Thus, the carrier exhibits transport characteristics similar to particles of 1 μm diameter but at a fraction of the pressure drop when operated in a packed bed.

Fibres are often operated in the form of hollow-fibre units. In this case, high flow rates and high protein or cell concentrations can be employed (Inloes *et al.* 1983). Hollow-fibre units are routinely used for medical applications such as dialysis and thus are commercially available in various forms and dimensions. The fibre material can be chosen to give the characteristics desired, such as molecular weight cutoff or hydrophobicity/hydrophilicity. Non-porous fibres have also been used in a packed bed reactor as an immobilisation support (Wikström and Larsson 1987).

Membranes have the advantage that extremely high flow rates can be achieved due to the high surface area to volume ratio when operated in a cross-flow or dead-end flow reactor. Flow is often driven across the membrane so that mass transport is driven by convection and not diffusion thereby guaranteeing rapid kinetics of interaction (Unarska *et al.* 1990). When flow is parallel to the membrane and mass transport is driven by diffusion, kinetics will also be relatively fast due to the short diffusion distance (thinness of the membrane). Probably the most formidable advantage of using membranes is the fact that a vast array of various membranes with innumerable characteristics are commercially available. Membranes are routinely used in molecular biology and biochemistry laboratories in such techniques as southern, northern and western blotting (Ausubel *et al.* 1990).

Benefits and Drawbacks

The main advantage that immobilisation procures is that it simplifies separation and recovery of the immobilised macromolecules or cell as well as any specific binding partner. The immobilised preparation can then be reused either in batch or in a continuous system and hence diminishes the cost of the process. For immobilised cell systems, for instance, dilution rates which far exceed the growth rate of the cells can be used without risk for cell washout as it would occur in the comparable free cell system. Immobilisation often enhances the stability of macromolecules or cells, for example, through multipoint attachment resulting in conformational fixation or by protecting the immobilised protein or cell from adverse conditions (Birnbaum *et al.* 1983; Martinek and Torchlin 1988; Monsan and Combes 1988). As a result the operating lifetime of the immobilised protein or cell will often be extended. The support material may also alter various parameters for optimal operation of the immobilised entity which may be advantageous i.e. pH, temp. Partitioning effects for various compounds may occur between the bulk and immobilised phases. Mass transfer limitation is often exhibited for immobilised systems, particularly those involving entrapment, and in some cases this may be advantageous, for example, when substrate inhibition occurs.

In many cases, the biochemical engineer must also decide whether to use a purified enzyme or whole cells when preparing the immobilised biocatalyst. In many instances, it may be advantageous to use the purified enzymes for example when side-reactions are to be a minimum. However, enzyme purification is tedious and costly and may not be cost effective for the process as a whole. Consequently, the use of whole cells may be superiour. Several other consideration must be taken into account when deciding whether to use enzymes or cells (Birnbaum *et al.* 1986).

Of course, immobilisation can also have detrimental effects. The immobilisation procedure used may destroy or significantly alter the protein or cell functionality. Alterations in pH, partitioning effects or mass transfer limitation can have negative effects. In many cases these can be alleviated. For example, negative mass transfer effects with respect to substrate, can be minimised by increasing agitation, decreasing particle size or increasing pore size.

Applications

Adsorption

Probably the most widely used immobilisation application is the enzyme linked immunosorbent assay. In this case, as we have used for the determination of γ-interferon (γ-IFN), monoclonal antibodies are adsorbed, presumably through hydrophobic interaction, to the non-porous bottom of 96 - well microtiter plates (Nilsson et al. 1988b). After blocking of remaining binding sites with an inert protein such as bovine serum albumin (BSA), the sample or standard is added, allowed to incubate (at which time the specific immunoaffinity binding occurs) and washed. A second polyclonal antibody directed against γ-interferon is then added, again allowed to interact and then washed. Finally, a secondary enzyme labelled antibody directed against the polyclonal antibody is added, incubated and washed. A colorimetric enzyme substrate is then added and after a period of time absorbance measurements are made which correspond to antigen concentration. In this case, both an initial hydrophobic interaction between the polystyrene plate and the primary adsorbed antibody and subsequent bioaffinity interactions between the antigen and the antibodies occur.

Another example of bioaffinity adsorption is the binding of animal cells to gelatin microcarriers which requires an initial binding of fibronectin to the support (Nilsson and Mosbach 1980). Anchorage dependent animal cell culture is often employed for the production of proteins of therapeutic interest, particularly glycosylated proteins which cannot be produced in procaryotic cells, such as γ-IFN or tissue-type plasminogen activator (tPA) (Nilsson et al. 1988b). Bioaffinity adsorption methods which involve protein-protein interactions are susceptible to destruction when proteases are present or produced in the immobilised system. This occurs when anchorage dependent recombinant Chinese Hamster Ovary cells are cultured for the production of tPA. In this case, poor cell attachment as well as low tPA yield is observed. Presumably the tPA produced converts the plasminogen present in the culture media to plasmin, a broad spectrum protease, which in turn proteolyzes the proteins required for cell adhesion and spreading (i.e. fibronectin). By adding a protease inhibitor such as aprotinin or ε-aminocaproic acid this effect can be reversed and thereby give higher cell as well as tPA yields.

Covalent Coupling

Various coupling techniques may be employed (Scouten 1987). One method is that based on preactivation of hydroxyl supports (such as agarose and silica) with sulfonyl chlorides such as tosyl and tresyl chloride (Nilsson and Mosbach 1988). We have used this method to construct an analytical procedure based on flow injection thermometric

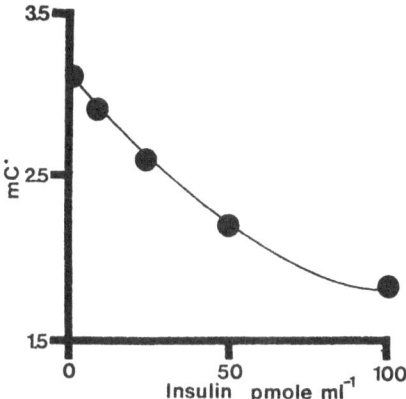

Fig. 2.2. Standard curve for flow injection thermometric enzyme linked immunosorbent assay of insulin.

enzyme linked immunosorbent assay, for the determination of human proinsulin produced by recombinant *E. coli* (Birnbaum *et al.* 1986). Thus, polyclonal antibodies directed against insulin were covalently coupled to tresyl chloride activated Sepharose, packed into a 1 ml column and placed into an enzyme thermistor unit. Sample was mixed with an insulin-peroxidase conjugate and automatically injected into the thermistor unit. Enzyme substrate was subsequently injected and followed by a 1 min. glycine wash to regenerate the immunosorbent for subsequent assay. The signal from the thermistor unit was converted with a wheatstone bridge with output on a recorder. A typical standard curve, measured against insulin, is shown in Fig. 2.2.

Table 2.1. Retention times, k', for various secondary alcohols and ketones with a column containing immobilised HLADH

Compound	Retention time, k'
Terbutaline	0.79
Bambuterol	1.29
Terbutaline ketone	2.93
Bambuterone	4.36

Another analytical application in which we have utilised tresyl chloride activation as a means of protein immobilisation is for the analytical separation of alcohols and ketones (Birnbaum and Nilsson 1991). Thus, horse liver alcohol dehydrogenase (HLADH) was coupled to tresyl chloride activated silica (pore size 300 Å) at a concentration of 100 mg per gram silica. The silica immobilised protein was packed into a stainless steel column (4.6 mm x 50 mm) and used for the isocratic separation of ß-blockers and their ketone derivatives with k' values as indicated in Table 2.1. This type of separation represents a weak affinity version of high performance liquid affinity chromatography (Zopf and Ohlson 1990; Ohlson *et al.* 1989).

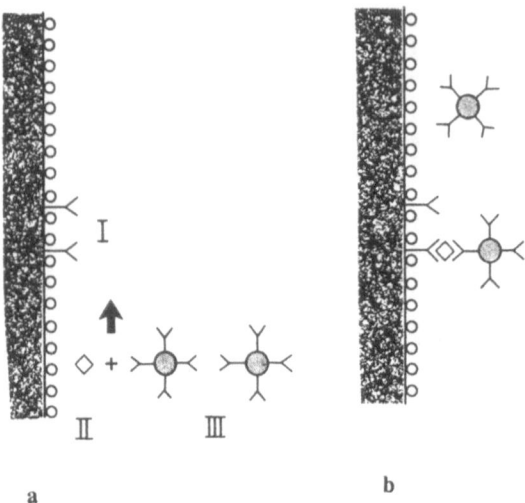

a   b

Fig. 2.3. Schematic representation of latex based thin layer immunoassay. I membrane immobilised monoclonal antibody II protein ie c-reactive protein III latex immobilised monoclonal antibody.

Other covalent coupling methods include the carbodiimide and the trichloro-s-triazine methods which we have employed to produce a latex based thin layer immuno-chromatographic method for protein quantitation (Birnbaum *et al.* submitted 1992). This method is based on the immunocapillarymigration principles described earlier and is shown schematically in Fig. 2.3 (Glad and Grubb 1978). It is based on a sandwich assay format and involves monoclonal antibodies of two distinct specificities for the protein to be analysed. One of the monoclonal antibodies is covalently immobilised to dyed latex particles of 0.3 µm diameter using carbodiimide as coupling reagent. The other monoclonal antibody is immobilised to a distinct region of a trichloro-s-triazine preactivated nylon membrane with pore dimensions of 3 µm. Antibody deposition on the preactivated membrane is achieved either by slot-blotting, stamping or with a piezoelectric driven ink-jet (Kimura *et al.* 1988) as used in a Siemens ink-jet printer PT 89 S (Birnbaum *et al.* unpublished). The assay is performed as follows: the sample is mixed and incubated five minutes with the antibody coated latex particles which are then allowed to migrate up the porous membrane. The antigen, now labelled with the dyed latex particles will bind to the membrane immobilised antibody. The intensity of colour which results from the antigen-antibody latex binding at the membrane immobilised antibody region is measured with a reflectometer. Antigen concentration is logarithmetrically proportional to the absorbance measurements as shown in Fig. 2.4. This method has been used for the quantitation of chorionic gonadotropin and c-reactive protein.

Another popular method for covalent coupling of proteins is the use of glutaraldehyde. We have used glutaraldehyde to cross-link BSA within fused silica capillaries of 75 µm diameter. Conditions were chosen so that gelation occurred a few minutes after BSA and glutaraldehyde were mixed. Hence, the mixture could be pumped into the capillary before the gel formed. The gel-filled capillary could then be used in a capillary electrophoresis instrument to separate the optical isomers of tryptophan (Birnbaum and Nilsson, submitted 1992).

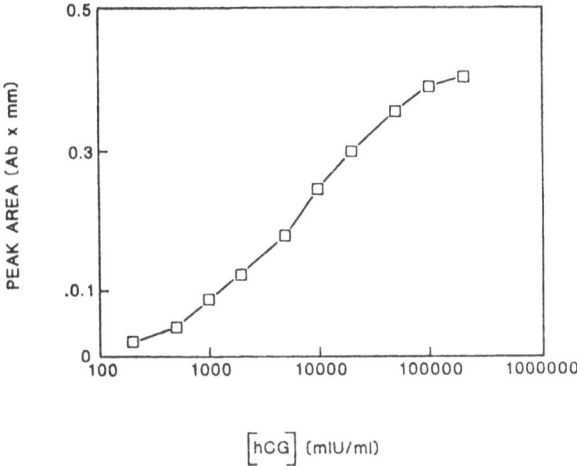

Fig. 2.4. Standard curve for chorionic gonadotropin using latex thin layer immunoassay.

Entrapment

Another immobilisation method which we frequently employ, in particular, for cell immobilisation, is entrapment. Among the various polymers used, cell entrapment in calcium alginate, is one of the more frequent methods. Alginic acid is a uronic polysaccharide which forms gels in the presence of certain cations such as calcium, strontium and barium. Calcium alginate beads are formed by dripping a sodium alginate-cell mixture into a calcium chloride solution. Ethanol production with calcium alginate immobilised cells is one of the more popular subjects as seen from its propensity in the literature.

Earlier, we investigated the possibility of co-immobilising magnetic material with entrapped yeast to facilitate separation of the immobilised preparation from the bulk phase by applying a magnetic field (Birnbaum and Larsson 1982). This type of "selective" retrieval was foreseen to be particularly advantageous when particulate solutions, such as partially hydrolysed biomass, were to be further fermented. We showed that the inclusion of 1 % (w/v) magnetite with calcium alginate entrapped *S. cerevisiae* had no effect on the fermentation activity nor on the viability of the immobilised cells.

Calcium alginate entrapment has the disadvantage that the matrix is unstable in media which complex calcium such as phosphate. Other cations, such as strontium, may be used to improve stability. We have shown that calcium alginate can be stabilised by covalently cross-linking the gel with polyethyleneimine without significantly affecting the catalytic activity or viability of the entrapped yeast cells (Birnbaum *et al.* 1981). In addition, we found that after treatment the mechanical properties of the preparation improved drastically as shown in Fig. 2.5. This stabilisation method has also been used to improve the stability of calcium pectate gel beads (P. Gemiener, 1991, personal communication).

As discussed above, one of the priorities for preparation of optimal immobilised cell beads is to form particles of small dimensions. Jet stream preparation of small

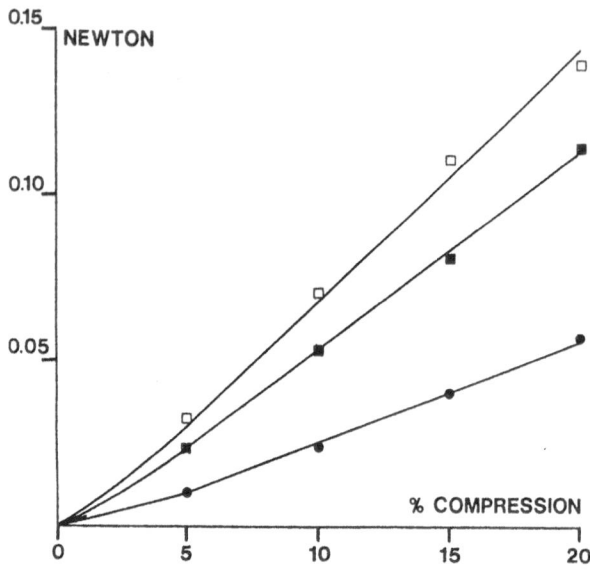

Fig. 2.5. Mechanical stability of calcium alginate beads treated with polyethyl-eneimine and glutaraldehyde (squares) compared to non-treated beads (circles).

spherical immobilised cell preparations has been reported (Birnbaum *et al.* 1983). We developed a general method for entrapping cells with preserved viability in beads of desired dimension (Nilsson *et al.* 1983). The method is shown schematically in Fig. 2.6. This method could be used with microbial, algal, plant and animal cells and in various support materials such as agar, agarose, alginate, carrageenan, fibrin and polyacrylamide. We later used this method for entrapment of recombinant *B. subtilis* and *E. coli* cells for the production of human proinsulin (Mosbach *et al.* 1983; Birnbaum *et al.* 1988).

Future Trends

It is inevitable that research and development of immobilised macromolecules and cells and their applications will continue. In particular we foresee that progress will occur in conjunction with developments in genetic engineering. For example, new or improved immobilisation methods and applications will be obtained through the use of site-directed mutagenesis, genetic fusion and recombinat cells. Other areas of future developments with immobilised macromolecules and cells include miniaturisation, particularly for analytical purposes. Other developments will focus on advances in support structure and preparation such as the use of continuous gels as well as the develpment of synthetic protein mimics through the imprinting technique (Hjertén *et al.* 1989; Ekberg and Mosbach 1989).

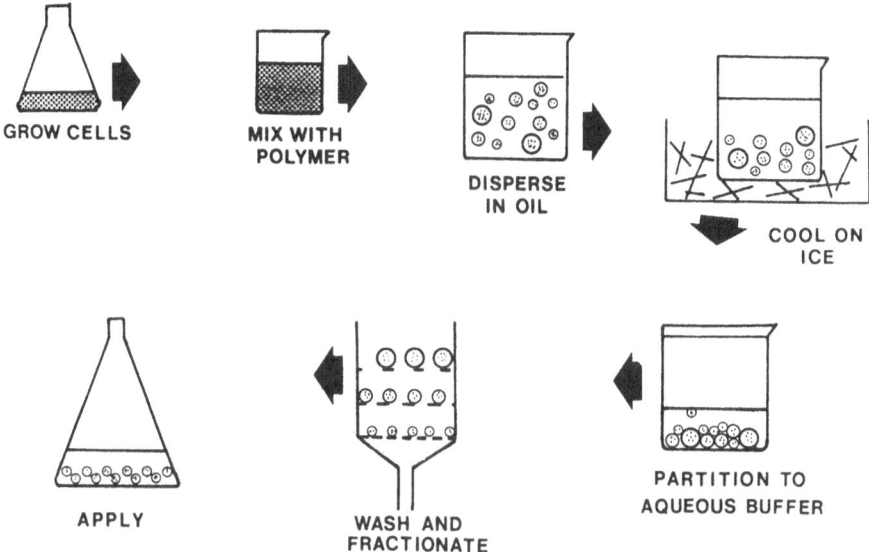

Fig. 2.6. General steps for entrapping viable cells in small spherical particles.

References

Afeyan NB, Gordon NF, Mazsaroff I, Varady L, Fulton SP, Yang YB, Regnier FE (1990) Flow-through particles for high-performance liquid chromatographic separation of biomolecules: perfusion chromatography. J. Chromatogr 519:1-29

Ausubel FM, Brent R, Kingston RE, Moore DD, Seidman JG, Smith JA, Struhl K (1990) Current protocols in molecular biology, Wiley, New York

Baranov VI, Morosov IY, Ortlepp SA, Spirin AS (1989) Gene expression in a cell-free system on the preparative scale. Gene 84:463-466

Birnbaum S and Larsson PO (1982) Application of magnetic immobilised microorganisms: ethanol production by Saccharomyces cerevisiae. Appl Biochem Biotechnol 7:55-57

Birnbaum S and Mosbach K (1991) Perspectives on immobilised proteins. Curr Opin Biotechnol 2:44-51

Birnbaum S and Nilsson K (1991) Dehydrogenase-silica as a stationary phase for the separation of alcohols and ketones. J Chromatogr 587:268-270

Birnbaum S and Nilsson S (1992) Protein-based capillary gel electrophoresis for the separation of optical isomers. Anal Chem submitted

Birnbaum S, Pendleton R, Larsson PO, and Mosbach K (1981) Covalent stabilisation of alginate gel for the entrapment of living whole cells. Biotechnol Lett 3:393-400

Birnbaum S, Larsson PO, Mosbach K (1983) Immobilised cells. In: Scouten WH (ed) Solid phase biochemistry, analytical and synthetic applications, Wiley, New York, pp 679-762

Birnbaum S, Larsson PO, Mosback K (1986) Immobilised biocatalysts: the choice bewieen enzymes and cells. In: Webb C, Black GM, Atkinson B (eds) Process engineering aspects of immobilised cell systems, The institution of chemical engineers, Rugby, England, pp 35-59

Birnbaum S, Bülow L, Hardy K, Danielsson B, Mosbach K (1986) Automated thermometric enzyme immunoassay of human proinsulin produced by Escherichia coli. Anal Biochem 158:12-19

Birnbaum S, Bülow L, Hardy K, Mosbach K (1988) Production and release of human proinsulin by recombinant Escherichia coli immobilised in agarose microbeads. Enzyme Microb Technol 10:601-605

Birnbaum S, Udén C, Magnusson CG, Nilsson S (1992) Latex based thin layer immunoaffinity chromatography for quantitation of protein analytes. Anal Biochem submitted

Braun S, Rappoport S, Zusman R, Avnir D, Ottolenghi M (1991) Biochemical active sol-gel glasses: the trapping of enzymes. Matt Lett 10:1-5

Coleman PL, Walker MM, Milbrath DS, Stauffer DM, Rasmussen JK, Krepski LR, Heilmann SM (1990) Immobilisation of protein A at high density on azlactone-functional polymeric beads and their use in affinity chromatography. J Chromatogr 512:345-363

Ekberg B and Mosbach K (1989) Molecular imprinting: a technique for producing specific separation materials. Trends Biotechnol 7:92-96

Glad C and Grubb A (1978) Immunocapillarymigration - a new method for immunochemical quantitation. Anal Biochem 85:180-187

Glad M and Larsson PO (1991) New methods for separation and recovery of biomolecules. Curr Opin Biotechnol 2.413-418

Glad M, Norrlöw O, Sellergren B, Siegbahn N, Mosbach K (1985) Use of silane monomers for molecular imprinting and enzyme entrapment in polysiloxane-coated porous silica. J Chromatogr 347:11-23

Hjertén S, Liao JL, Zhang R (1989) High-performance liquid chromatography on continuous polymer beds. J Chromatogr 473:273-275

Inloes DS, Taylor DP, Cohen SN, Michaels AS, Robertson CR (1983) Ethanol production by *Saccharomyces cerevisiae* immobilised in hollow-fiber membrane bioreactors. Appl Environ Microbiol 46:264-278

Kimura J, Kawana Y, Kuriyama T (1988) An immobilised enzyme membrane fabrication method using an ink jet nozzle. Biosensors 4:41-52

Kobos RK, Eveleigh JW, Arentzen R (1989) A novel fluorocarbon-based immobilisation technology. Trends Biotechnol 7:101-105

Martinek K and Torchilin VP (1988) Stabilisation of enzymes by intramolecular cross-linking using bifunctional reagents. Meth Enzymol 137:615-626

Monsan P and Combes D (1988) Enzyme stabilisation by immobilisation. Meth Enzymol 137:584-598

Mosbach K (1987a) Immobilised enzymes and cells, part b, Academic Press, Orlando (Methods in enzymology, vol 135)

Mosbach K (1987b) Immobilised enzymes and cells, part c, Academic Press, Orlando (Methods in enzymology, vol 136)

Mosbach K (1988) Immobilised enzymes and cells, part d, Academic Press, Orlando (Methods in enzymology, vol 137)

Mosbach K and Mosbach R (1966) Entrapment of enzymes and microorganisms in synthetic cross-linked polymers and their application in column techniques. Acta Chem Scand 20:2807-2810

Mosbach K, Birnbaum S, Hardy K; Davies J, Bülow L (1983) Formation of proinsulin by immobilised *Bacillus subtilis*. Nature 302:543-545

Nilsson K and Mosbach K (1980) Preparation of Immobilised Animal Cells. FEBS Lett 118:145-150

Nilsson K and Mosbach K (1987) Tresyl chloride-activated supports for enzyme immobilisation. Meth Enzymol 135:65-78

Nilsson K, Birnbaum S, Flygare S, Linse L, Schröder U, Jeppsson U, Larsson PO, Mosbach K, Brodelius P (1983) A general method for the immobilisation of cells with preserved viability. Eur J Appl Microbiol Biotechnol 17:319-326

Nilsson K, Buzsaky F, Mosbach K (1986) Growth of anchorage-dependent cells on macroporous microcarriers. Bio/technology 4:989-990

Nilsson K, Buzsaky F, Birnbaum S, Mosbach K (1988) Macroporous microcarriers. In: Spier RE, Griffiths JB (eds) Modern approaches to animal cell technology, Butterworth, Seven Oaks, pp 492-503

Ohlson S, Hansson L, Glad M, Mosbach K, Larsson PO (1989) High performance liquid affinity chromatography: a new tool in biotechnology. Trends Biotechnol 7:179-186

Olson CW, Leung SK, Yarmush ML (1989) Recovery of antigens from immunoadsorbents using high pressure. Bio/technology 7:369-373

Powell LW (1991) Immobilised biocatalyst technolgy. In: Microbial technology, Elsevier, Amsterdam pp369-394

Salter GJ and Kell DB (1991) New materials and technology for cell immobilisation. Curr Opin Biotechnol 2:385-389

Scouten WH (1987) A survey of enzyme coupling techniques. Meth Enzymol 135:30-65

Scouten WH (1991) Affinity chromatography for protein isolation. Curr Opin Biotechnol 2:37-43

Stewart DJ, Purvis DR, Lowe CR (1990) Affinity chromatography on novel perfluorocarbon supports: immobilisation of C.I. Reactive Blue 2 on a polyvinyl alcohol-coated perfluoropolymer support and its application in affinity chromatography. J Chromatogr 510:177-187

Sun A (1988) Microencapsulation of pancreatic islet cells: a bioartificial endocrine pancreas. Meth Enzymol 137:575-580

Taylor RF (1991) Protein immobilisation, fundamentals and applications, Marcel Dekker, New York (Bioprocess technology, vol 14)

Unarska M, Davies PA, Esnouf MP, Bellhouse BJ (1990) Comparative study of reaction kinetics in membrane and agarose bead affinity systems. J Chromatogr 519.53-67

Wikström P and Larsson PO (1987) Affinity fibre - a new support for rapid enzyme purifcation by high-performance liquid affinity chromatography. J Chromatrogr 388:123-134

Zopf D and Ohlson S (1990) Weak-affinity chromatography. Nature 346:87-88

Chapter 3

Immunoaffinity Chromatography and Oriented Immobilisation of Antibodies

J. Káš, J. Sajdok and F. Strejček

Introduction

Affinity chromatography (also bioaffinity or biospecific chromatography), in general, is based on the ability of various biologically active substances to bind specifically and reversibly other substances. One of the interacting counter-parts is attached, usually by covalent coupling, to a suitable solid support and it is called affinant or affinity ligand. (For details see, for instance, Turková 1978, Scouten 1981, Dean 1985).

Immunoaffinity chromatography (IAC) represents a special case of affinity chromatography where a specific noncovalent interaction occurs between an antigen and corresponding antibody. Any of the mentioned counter-parts may act as an affinity ligand. Here, the immobilisation of an antibody and separation of the antigen will be presented as an example. Such an arrangment of immunoaffinity chromatography includes the adsorption of an antigen (or hapten) present in the applied sample on the immunosorbent (i.e. to antibody bound to a solid support), subsequent washing with a suitable buffer enabling the removal of unspecifically adsorbed proteins and elution of the antigen (or hapten) from the column by an appropriate elution reagent (Fig. 3.1). The successful performance of IAC requires careful optimisation and depends on many factors which are summarised in Figure 3.2. The crucial problem of IAC is the preparation of an appropriate antibody (either polyclonal or monoclonal) with a suitable avidity (or affinity), stability, specificity and the possibility of preparation in a reproducible quality and sufficient quantity. The second step is the preparation of a good immunosorbent with high immobilisation yield, proper ligand orientation, maintenance of immunoreactivity, sufficient stability (no leakage of antibodies) and satisfactory flux properties. The performance conditions of IAC must take account of the stability of antigen and antibody during adsorption and desorption and eliminate the effect of proteinases if present in the sample. Finally, the economic factors (shelf-life of the column, separation yields etc.) determine the practical use of the proposed IAC.

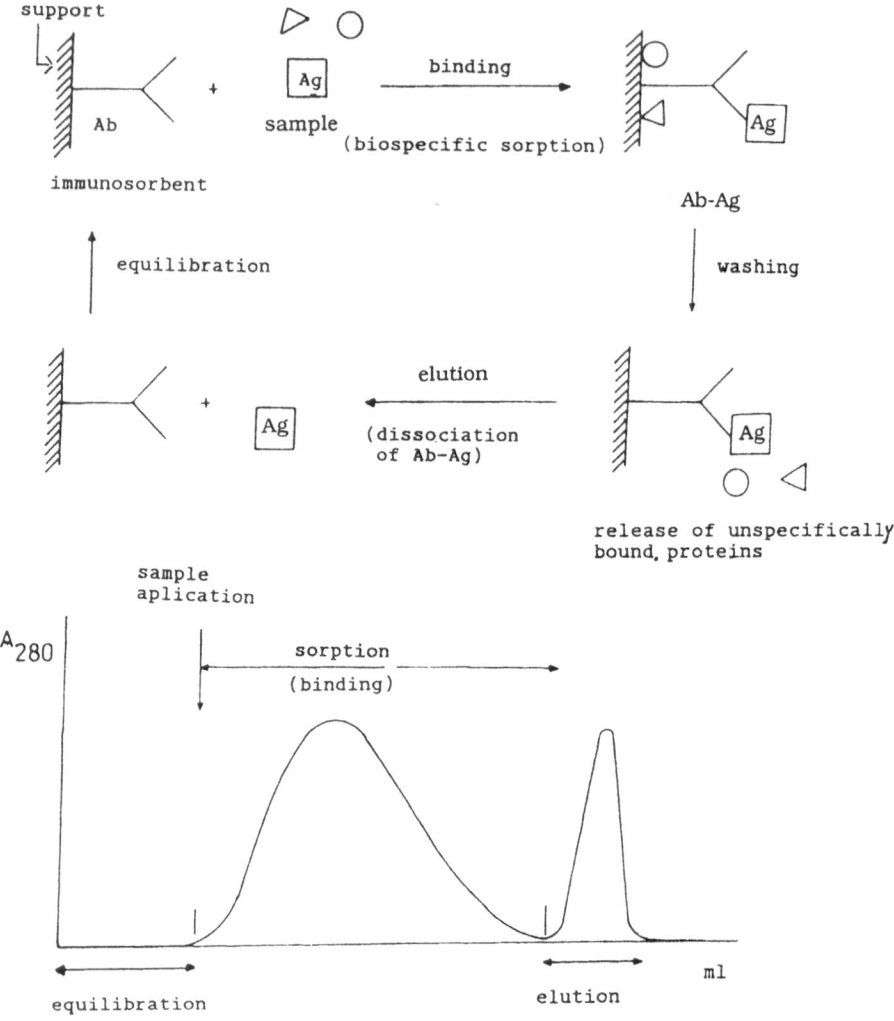

Fig. 3.1. Separation of an antigen from the sample on the immobilised antibody.

Antibodies for IAC

High specificity and low avidity (affinity) are in general required for IAC (for polyclonal antibodies as well as for monoclonal antibodies). Low avidity (affinity) of antibody makes possible antigen desorption under gentle conditions (Ruoslahti 1976). This requirement is in contrast with demands of immunoassays. The low avidity of polyclonal antibodies may be achieved using the following approaches.

General Schema of IAC

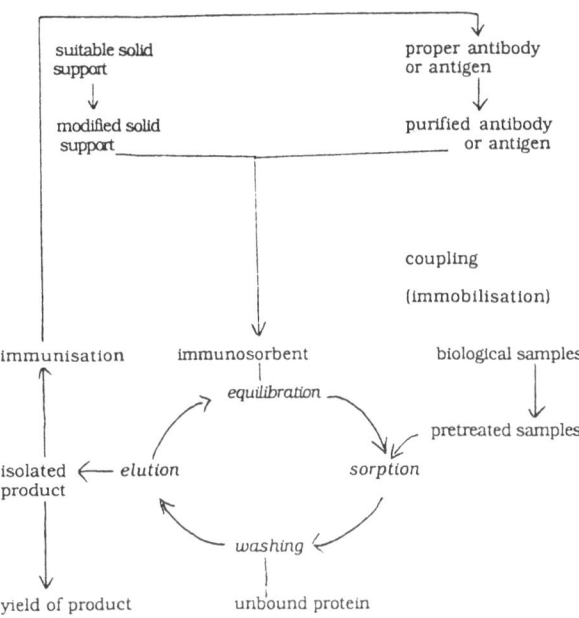

Fig. 3.2. Factors affecting the performance of IAC.

Selection of the Immunisation Conditions

The antisera obtained in an early stage after the first immunisation usually show lower avidity than antisera obtained after several immunisations and longer immunisation periods (Werblin *et al.* 1973). The disadvantage of such antisera is the low content of specific antibodies. High immunisation doses of an antigen induce the formation of antibodies with low avidity, however, the probability of simultaneous creation of antibodies against contaminating proteins is increased. Antisera with low specificity may be formed.

Application of Homologous Proteins

Antibodies may show specific interaction with homologous proteins. Such interaction is usually less firm than with the antigen used for immunisation. Erikson and Sandman (1977) used rabbit antibody against N-acetyl-ß-glucosamidase (NAG) from human liver for IAC of various NAGs. Rabbit, sheep and bovine NAG were not attached to an immunosorbent while rat and mouse NAG were bound and subsequently were easily eluted from the immunosorbent under very gentle conditions (1 M NaCl). These antigens, however, did not show any immunoprecipitation with rabbit antibody against NAG. The capacity of the immunosorbent was similar for human, rat and mouse NAG when buffer pH 2.6 was used for elution (Erikson and Sandman 1977). Likewise the antibodies against lethal factor from the venom of *Physaliaphysalis* bound to Sepharose 4B-CN were used for the isolation of the lethal factor from *Vespa orientalis* (Russo *et al.* 1983). This antigen was eluted by a simple increase of the ionic strengh of the phosphate buffer pH 7. ß-Galactosidase from *Aerobacter cloacae* was eluted from the immunosorbent containing bound antibodies against ß-galactosidase from *Escherichia coli* only by the adsorption buffer (Erikson and Steers 1970).

Separation of Antibody Subpopulations on the Sorbent with Immobilised Antigen

A number of antibody subpopulations mutually differing in their affinity to the antigen are created in the blood of the immunised animal. They may be separated on the sorbent with bound antigen by means of programmed elution. Folkersen *et al.* 1985 eluted three antibody populations against α-1-fetoprotein by means of a stepwise elution. The first antibody subpopulation was eluted with 1 M NaCl, the second one with 0.1 M acetate-HCl pH 3 and the third one with 0.1 M acetate-HCl pH 2.3. They immobilised each of these three antibody subpopulations on Sepharose 4B-CN and used the prepared immunosorbents for the isolation of α-1-fetoprotein. The immobilised antibody of the first subpopulation bound the antigen very weakly, 38% of the antigen was eluted already during the immunosorbent washing. Using the immunosorbent with the second antibody subpopulation the amount of released antigen during washing decreased to 12%, 33% of the antigen was then eluted at pH 3.3 and the rest at pH 2.3. The third antibody subpopulation bound the antigen very firmly and only 50% was eluted at pH 2.3. A similar approach was applied for the isolation of IgG subpopulations in pH gradient with addition of acetonitrile.

Chemical Modification of Immobilised Antibodies and Antigens

The chemical modification of the immobilised antibodies may decrease their avidity. Likewise the application of the chemically modified antigen for the immunisation may either decrease or increase the avidity of received antibodies (Ruoslahti 1978).

Use of Monoclonal Antibodies

During the preparation of monoclonal antibody clones showing low affinity (i.e. having dissociation constant higher than $10^{-4}M$) may be selected. Such antibodies, after their immobilisation, are well suited for immunoaffinity chromatography (Ohlson 1988).

Immobilisation of Antibodies to Solid Supports

Immobilisation yields, sufficient steric availability of the binding sites of antibodies to antigens after immobilisation, maintenance of immunoreactivity and a sufficiently stable antibody solid-support linkage at conditions used for antigen elution are the main criteria for the selection of the proper immobilisation technique. In contrast to immunoassays, covalent binding of antibody is almost exclusively used either by using the properties of functional groups of protein part or by utilising original or modified carbohydrate moieties of immunoglobulins.

Immobilisation of Antibodies via Functional Groups of their Protein Parts

One of the oldest methods of immunosorbent preparation is the crosslinking of immunoglobulins with glutaraldehyde or with ethylchloroformate. This procedure leads, however, to the formation of particles with unequal size and shape and unsatisfactory flux properties (Ternynck and Avrameas 1969). The most frequent method applied is the immobilisation of immunoglobulins on Sepharose 4B or 2B activated with bromcyan (Axén and Ernbrac 1967). The toxic properties of the activation agent, unstability of the activated Sepharose and lack of spacer are disadvantages of this procedure. In spite of these drawbacks the method has been studied in detail by many authors. The immobilisation of immunoglobulins is often recommended at pH 8.7 - pH 9.5. The reaction under these conditions favours the multipoint antibody attachment via ε-amino groups and this may cause changes in the tertiary structure of immunoglobulins and the decrease in immunoreactivity. Cuatrecasas (1971) proved this assumption from the experiments with sheep immunoglobulins against bovine insulin and showed that the more suitable pH for immobilisation is 6.5. The immunoglobulins bound at pH 9.5 lost almost all immunoreactivity. As opposed to this, the application of a lower pH causes a decrease in the immobilisation yields (Axén and Ernback 1978). The high capacity immunosorbent has been prepared by periodate oxidation of cellulose suspension (400 mg - 1000 mg of IgG per gram support) gained by its precipitation from cupric hydroxide. This immunosorbent had also very good flux properties (Gurvich and Lechtzind 1982). Among the other widely applied immobilisation procedures, carbohydrates activated with halopyrimidines or with reactive azodyes, polyacrylhydrazidoagarose (Miron and Wilchek 1981), hydroxyalkylmethacrylate (Goffinet et al. 1975), aminohexyl-Sepharose activated with glutaraldehyde (Cambiaso et al. 1975), and agarose activated with divinylsulphate should be mentioned. Immunoadsorption film has been obtained by the polymerisation of a gelatin and polyethylendiamine mixture with glutaraldehyde. The glutaraldehyde in excess is then utilised for antibody binding (Bröcker and Sorg 1977). The antibodies were also immobilised by radiation polymerisation using hydroxyethylmethacrylate, neopenthylglycoldimethacrylate and trimethylpropantriacrylate (Kumakura et al. 1983). The selection of the appropriate immobilisation technique and suitable solid support depends on the conditions of IAC. For instance, Sepharose activated with divinylsulphate is more suitable for the immobilisation of an antibody than Sepharose activated with BrCN if alkaline buffers are supposed to be used for antibody elution. (Lilme et al. 1986).

Oriented Immobilisation of Antibodies

The immobilisation of antibodies via the Fc region of the heavy immunoglobulin chain seems to be very suitable because the antibody binding sites are located on the opposite part of the molecule. Likewise, the attachment via Fab fragments or reduced F(ab)$_2$ using SH groups also makes possible the correct orientation of antibodies bound to a solid support surface.

Immobilisation of Antibody by Using Protein A and Protein G

Protein A from *Staphylococcus aureus* interacts specifically with the Fc part of the immunoglobulin molecule (Forsgren and Sjöquist 1966, Lindmark 1982, Langone 1982).This noncovalent interaction may be subsequently stabilised by covalent crosslinking with various bifunctional reagents, e.g. by glutaraldehyde. Pure protein A is able to bind two molecules of IgG per molecule (Hjelm *et al.* 1975). Immobilised protein G (e.g. protein G agarose from Calbiochem AG) binds all IgG subclasses, but not IgA and IgM and it may be used for IgG purification (Boyle and Reis 1987). At least five various proteins from *Streptococci* and *Staphylococci* are now known which react in similar way (Myhre and Kronvall 1981).

a

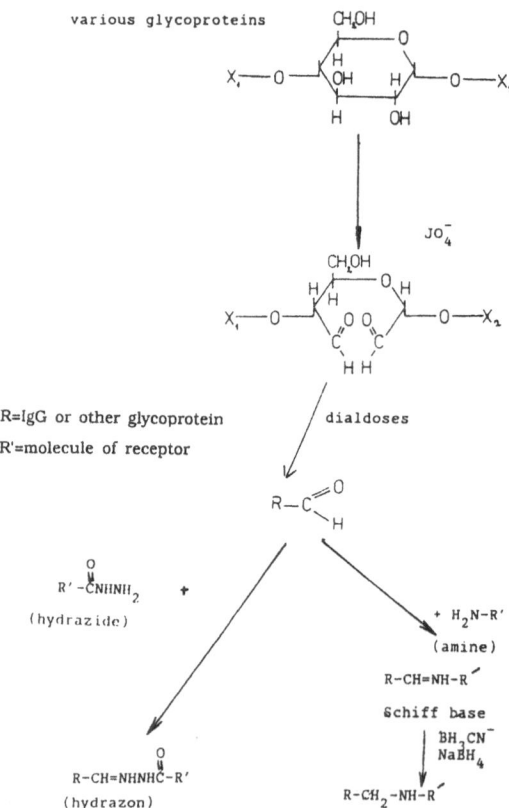

b

neuraminidase

$-(Fucose)_{0-1}- (Manose)_3-(Galactose)_{1-2}-(N-acetylglucosamine)_{3-4}-$ (sialic acid)$_{0-3}$

$-(Fucose)_{0-1}- (Manose)_3-(Galactose)_{1-2}-(N-acetylglucosamine)_{3-4}$

galactose oxidase (GaO)

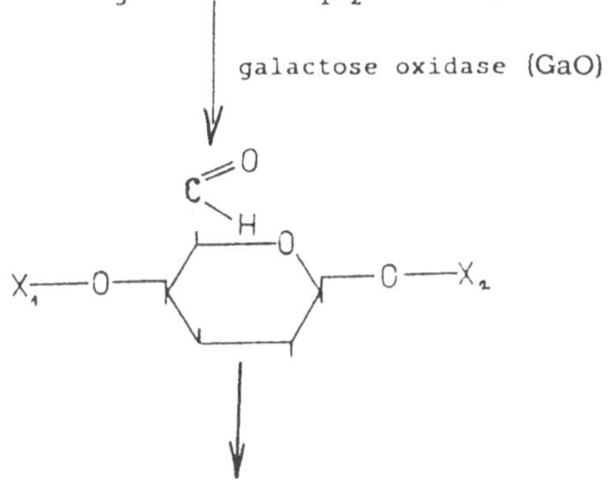

further procedure as given in Figure 3.3a

Fig. 3.3 a Immobilisation of an antibody after chemical oxidation of its carbohydrate moiety. **b** Immobilisation of an antibody after enzymatic oxidation of its carbohydrate moiety.

Immobilisation of Antibodies via the Carbohydrate Moiety of IgG

The major part of the carbohydrate moiety of IgG is located in its Fc region. Thus it may be expected that its attachment to the solid support via this place will be sterically advantageous (O'Shannessy and Quarles 1987). The antibodies immobilised via the carbohydrate moiety showed remarkably higher immunoreactivity than those immobilised to Sepharose activated with BrCN or to Affi-gel. For instance, the ratio between the bound trypsin inhibitor and polyclonal antibody immobilised via its carbohydrate moiety was 1.4, while the usual ratio after the usage of unoriented immobilisation techniques is 0.4-0.6 (Hoffman and O'Shannessy 1988). In our analogue experiments with ovalbumin we have reached an antigen/antibody ratio of 1.7 (Turková et al. 1991). The theoretical ratio between bound antigen and an immobilised IgG should be 2.0, because IgG contains two binding sites for antigens. Also the experiments of Prisažny et al. 1988 confirmed that the oriented immobilisation of immunoglobulins brings sufficient sterical availability of the antibody binding sites, saves immunoreactivity and reduces antibody leakage from the solid support even at extreme conditions used for the elution of antigens.

Two main approaches may be applied in principal for the immobilisation of antibodies to solid supports via their carbohydrate parts.

The first approach utilises the reactivity of aldehyde groups formed by the oxidation of the carbohydrate part of IgG and subsequent successive reaction with the solid support actived with hydrazides or amines (Lenten and Aschell 1971) (Fig. 3.3). The oxidation of the carbohydrate part of IgG may be performed either chemically (e.g. with periodate) or enzymatically (Baumstark *et al.* 1964). For the enzymatic oxidation of galactose occuring in the carbohydrate domain of IgG galactose oxidase (E.C. 1.1.3.9) may be applied. Galactose oxidase oxidises hydroxyl groups in position C_6 of terminal galactose or N-acetylgalactoseamine to aldehyde. Solomon *et al.* 1989 have oxidised monoclonal antibodies by means of an immobilised complex of neuraminidase and galactose oxidase. The carbohydrate moiety of the antibody which is located far away from the hypervariable region contains the following residues: zero or one fucose, three mannoses, one or two galactoses, three or four N-acetylglucosamines and up to three sialic acids. The sialic acid, as a terminal residue, may be removed by neuraminidase and one galactose residue will become accessible to oxidation by galactose oxidase. The immunoreactivity of monoclonal antibodies immobilised after enzymatic oxidation was much better than those oxidised by periodate.

The aldehyde groups of the IgG formed by enzymatic or chemical oxidation may be coupled either with amines or hydrazides. The reaction with amines yields labile Schiff's bases which may form undesirable intra-molecular cross-linking between amino groups of the immobilised protein and aldehyde groups of its carbohydrate moiety.

The Schiff's base stabilisation either with sodium borohydride or preferentialy with sodium cyanoborohydride may have a deleterious effect on the immunoreactivity of bound immunoglobulin.

The condensation between aldehyde groups of the oxidised glycoprotein and added hydrazides creates much more stable hydrazones and eliminates the formation of intramolecular cross-linking. This is primarily caused by the very low pK of hydrazides, usually around 2.6 compared with primary amines which have pK's in the range of nine to ten. When the reaction takes place at pH 4.5-5.5 the majority of free amino groups of the immobilised protein are protonated and thus unreactive. This is the reason why the reaction with hydrazide modified supports is preferred to supports with amino groups.

The second approach is based on the interaction of the carbohydrate part of IgG with lectins, most frequently with concanavalin A. Concanavalin A reacts specifically with unsubstituted hydroxyl groups in C_3, C_4 and C_6 positions of α-D-mannopyranoside and α-D-glucopyranoside residues located at the nonreducible end of polysaccharides and glycoproteins (Ochoa 1981, Waseem and Salahuddin 1983). Lectins are immobilised on the solid support and the antibodies may be biospecifically adsorbed (Gersten and Marchalonis 1978). The attachment of the antibody may be fixed by subsequent crosslinking with bifunctional reagents (Husain and Salumuddin 1986).

Oriented Immobilisation of Antibodies by means of SH Groups of FaB Fragments

IgG may be split off specifically either with papain or with pepsin (Prisyazhnoy 1988). The fragments formed by pepsine cleavage are suitable for immobilisation via their SH groups formed after the preliminary reduction with 2-mercaptoethanol and $NaAsO_2$ (Brennan 1986).

Antigen Sorption to the Immunosorbent

The immunosorption is affected by the properties of the immunosorbent (depending on the properties of the solid support, quality of antibodies, immobilisation technique used, size and shape of the column), concentration of the antigen in the applied sample, temperature and pH kept during the sorption, salt concentration in the binding buffer and flow rate applied. The optimisation of the immunosorption on the already chosen immunosorbent is usually limited to searching of the proper binding buffer with a suitable salt concentration (pH usually applied is around 7-9), antigen concentration and flow rate. In general, antigen concentrations up to one half of column capacity will be satisfactory and flow rate may be used quite high. The most serious problem of IAC is often nonspecific sorption which depends on the composition of the sample and it may be partially limited by the way of column washing prior to antigen elution.

Antigen Desorption from the Immunosorbent

The desorption must be relatively gentle and it should not affect the properties of antigen and antibody. It must be undertaken immediately after the washing of the immunosorbent and as soon as possible after antigen binding.

The desorbed antigen must be immediately transferred to the environment protecting it against denaturation (adjustment of pH etc). The selected desorption procedure should allow the total desorption of the antigen. Alternatively, a combination of a few methods will cause the complete removing of the antigen. The desorption conditions must be considered in relation to the properties of the isolated antigen (its stability and losses of biological activity in different media etc) and the stability of immunosorbent from the viewpoint of long-term utilisation.

Antigen Elution by Means of pH Change

The antigen-antibody complex dissociates at extreme pH values. Low pH values are most frequently used (e.g. diluted HCl pH 1.8, glycine-HCl buffer pH 1.8-2.2, sodium citrate pH 2.6, sodium acetate pH 2.8). In spite of these conditions, it is possible to obtain many proteins in their native state. Ferrua and Masseyeff (1985) showed that the dissociation of antigen-antibody complexes only takes place in a narrow pH range. Some peptide bonds are broken in alkaline media and the antigen elution at alkaline pH is used only very rarely (Simons and Bearn 1969).

Application of Chaotropic Reagents

A systematic study of the dissociation of antigen-antibody complexes was published by Dandliker (1967). He found that the effectivity of desorption decreases in the following order: $SCN^- < ClO_4^- < I^- < Br^- < Cl^-$. Avrameas (1967) compared the dissociation effect of iodine, bromide and chloride ions in the concentration range 1.0 M-5.5 M on the immune complex of human IgG and pig anti IgG. He found 2 M NaI to be the best dissociation reagent. Its use was much more gentle than elution with glycine-HCl buffer at pH 2.2. The optimal yields of IAC may be achieved only at optimal concentration of the chaotropic reagent during the elution (Nishi and Hirai 1972). Attention should also be paid to the selection of the correct pH. According to our experiences a high shelf life of the immunosorbent may be achieved through the use of NH_4 SCN.

Hypotonic Antigen Desorption (Desorption with Decreasing Ionic Strength)

Ovalbumin, globulin from sunflower, α- and ß-amylase were eluted from the corresponding immunosorbents with distilled water giving a yield of about 50%. Subsequent desorption with sodium acetate pH 2.8 allows the removal of the residual proteins (Bureau and Daussant 1981,1983). Several authors (Ternyck and Avrameas 1971) have used distilled water for the elution of various antigens from immunosorbents. The antigen dissociation with distilled water may be sometimes improved with an interruption in elution (Bureau and Daussant 1983).

Desorption of Antigens with Guanidine and Urea

Guanidine and urea are often used for elution of antigens from immunosorbents. For instance, the antibodies against thyrotropin hormone bound to Sepharose 4B CN were applied for the separation of thyrotropin hormone, follicle stimulating hormone, luteinising hormone and growth hormone. After the binding to immunosorbent the individual hormones were partially separated in a guanidine gradient 0.1 M-4.0 M formed in the elution buffer guanidine-HCl (pH 3.2) at 4°C (Pekonen *et al.* 1980).

Elution of Antigens with Nonpolar Solvents

The addition of a nonpolar solvent to the elution buffer is effective in the dissociation process only when hydrophobic amino acid residues are present in the hypervariable part of IgG. Under such circumstances the addition of acetonitrile or similar nonpolar solvents facilitates the antigen desorption. For instance, the specific anti-human prolactin was separated by 20% acetonitrile in pH gradient 2-7. In contrast, minimal resolution was achieved in the absence of acetonitrile, immunoactivity was decreased and yields were very low (Lowry 1982).

Other Means of Elution

Except for the above mentioned modes of antigen elution from immunosorbents, various physical methods are under investigation, such as treatment with direct current or sonication (Bartoli and Roggero 1985).

Conclusions

IAC is undoubtedly a very versatile separation technique. Its attraction is in its universality derived from the general feature represented by the interaction between antigen and antibody. This universality involves many problems (e.g. oriented immobilisation of antibodies) the solution of which can be applied in other cases. On the other hand the variability of antibodies allows separation procedures for almost all low and high molecular weight compounds to be established. The possibilities for IAC are almost boundless. The only problem is to make IAC effective and economical for large scale applications.

The first successful industrial separations, mainly in the pharmaceutical industry, indicate that the era of IAC in practice has started.

Acknowledgement

The authors are obliged to Dr. Daussant (CNRS, Meudon, France) for fruitful discussions during his stay in Prague.

References

Axen R, Emback S (1971) Chemical fixation of enzymes to cyanogen halide activated polysacharide carriers. J Biochem. 18:351-360

Axen R, Porath J, Emback S (1967) Chemical coupling of protein to agarose. Nature 215:1491-1492

Avrameas S, Ternynk T (1967) Use of iodide salts in the isolation of antibodies and the dissolution of specific immune perecipitates. Biochem J 102:37-39

Baumstark JS, Laffin RJ, Bardwil WA (1964) A preparative method for the separation of 7S γ-globulin from human serum. Arch Biochem Biophys 108:514-522

Boyle MDP, Reis KJ (1987) Bacterial Fc receptors. BioTechnology 5:697-703

Bureau D, Daussant J (1981) Immunoaffinity chromatography of proteins a gentle and simple desorption J Immunol Meth.41:387-392

Bureau D, Daussant J (1983) Desorption following immunoaffinity chromatography, generalisation of a gentle procedure for desorbing antigen. J Immunol Methods 57:205-213

Brennan M (1986) A chemical technique for the preparation of biospecific antibodies from Fab' fragments of mouse monoclonal IgG$_1$ BioTechniques 5:424-427

Bröcker EB, Sorg C (1977) Specific separation of T-lymphocytes on immunoabsorptive films J Immunol Methods14:333-342

Cuatrecasas P, Anfinsen CB (1971) Affinity chromatography. Ann Rev Biochem 40:259-278

Dandliker WK, Sausure VA, Alonso R, Kierszenbaum F, Levison SA, Schapiro HC (1967) The effects of chaotropic ions on the dissociation

Dean PDG, Johnson WS, Midle FA (1985) Affinity Chromatography IRL Press Oxford England

Erickson RP, Sandman R (1977) Heterologous immunoadsorption in the purification of enzymes: N-acetyl-ß-glucosaminidase. Experientia 331:14-15

Erickson RP, Steers E (1970) Isoenzymes of bacterial ß-galactosidases:purification and characterisation of an isoenyzme forming ß-galactosidase of Aerobacter (Enterobacter) cloacae. Arch Biochem Biophys 137:399-408

Ferrua B, Masseyeff R (1985) Effect of pH on the recovery of human α-fetoprotein from immunosorbent columns. An improved elution procedure. J Immunol Methods 84:375-379

Folkersen J, Tiesner B, Westergaard JG, Grudzinskas JG (1985) The selection of antibodies with defined desorption properties from precipitated immune complexes for use in immunosorption procedures J Immunol Meth 77:44-54

Forgsgren A, Sjöquist J (1966) "Protein A" from S aureus.1 Pseudo-immune reaction with human γ-globulin. J Immunol 97:822-827

Gersten DM, Marchalonis JJ (1978) A rapid, novel method for the solid-phase derivatisation of IgG antibodies for immune-affinity chromatography. J Immunol. Methods 24:305

Goffinet A, Cambioaso GL, Vaerman JP, Heremans JF (1975) Glutaraldehyde-activates aminohexyl derivate of Sepharose 4B as a new versatile immunosorbent Immunochemistry 12:273-8

Gurvich AE, Lekhtsind EW (1982) High capacity immunoadsorbents based on preparations of reprecipitated cellulose. Mol Immunol 19(4):637-640

Hoffman WL, O Shannessy DJ (1988) Site-specific immobilisation of antibodies by their oligosaccharide moieties to new hydrazide derivatised solid supports. J Immunol Methods 112:113-120

Husain Q, Saleemuddin M (1986) Immobilisation of glycoenzymes using crude concavalin A and glutaraldehyde. Enzyme Microb Technol 8:686-690

Hjelm H, Sjödahl J, Sjöquist J (1975) Immunologically active and structurally similar fragments of protein A from Staphylococcus aureus. Eur J Biochem 57:395-403

Kumakura M, Kaetsu I, Suzuki M, Adachi S (1983) Immobilisation of antibodies and enzyme-labeled antibodies by radiation polymeration. Appl Biochem Biotech 8:87-96

Langone JJ (1982) Application of Immobilised Protein A in Immunochemical Techniques. J Immunol Methods 55:277-296

Van Lenten L, Ashwell G (1971) Studies on the chemical and enzymatic modification of glycoproteins. A general method for the tritiation of sialic acid-containing glycoproteins. J Biol Chem 246:1889-1894

Lindmark R (1982) Fixed protein A-containing Staphylococci as solid-phase immunoadsorbents. J Immunol. Methods 52:195-203

Lilme A, Nielsen CS, Larsen KP, Muller KG, Bog-Hansen TC (1986) Formation of stable and non-leaking affinity matrices using divinylsulphone-activated agarose following immobilisation of immunoglobulins and other proteins. J Chromatogr 376:299-305

Lowry PJ, Hodgkinson SC (1982) Selective elution of immuno- adsorbed anti-(human prolactin) immunoglobulins with enhanced immunochemical properties. Biochem J 205:535-541

Miron A, Wilchek M (1981) Poly(acrylhydrazido)agarose preparation via periodate oxidation and use for enzyme immobilisation and affinity chromatography. J Chromatogr 215:55-63

Myhre E, Kronvall G (1981) in:Basic Concepts of Streptococci and Streptococcal Diseases (Holm S, Chrestensen P, eds), Red book Ltd, Chertey UK p 209

Nishi S, Hirai H (1972) Purification of human, dog and rabbit α-fetoprotein by immunoadsorbents of Sepharose coupled with anti-human α-fetoprotein. Biochim Biophys Acta 278:293

Ochoa JL (1981) Consideration of the nature of the lectin-carbohydrate interaction. J Chromatogr. 215:351-360

Ohlson S, Landhlad A (1988) Novel approach to affinity chromatography using "weak" monoclonal antibody Analytical Biochemistry 169:204-208

Pekonen F, Williams D, Weintraub D (1980) Purification of thyrotropin and other glycoprotein hormones by immunoaffinity chromatography. Endocrinology 106:1327-1332

Prisyazhnoy VS, Fusek M, Alakhov YB (1988) Synthesis of high-capacity immunoaffinity sorbents with oriented immobilised immunoglobulins or their Fab fragments for isolation of proteins. J Chromatogr, 424:243-253

Ruoslahti E (1976) Antigen-antibody precipitates and immunoadsorbents. Applications to purification of α-fetoprotein. Scand J Immunol, Suppl 3 (Immunoadsorbents Protein Purification) 39-49

Ruoslahti E (1978) Immunochromatography on insolubilised antibodies of very low affinity:application to immunoadsorbence of bovine α-fetoprotein. J Immunol 121:1687-1690

Russo AJ, Cobbs CS, Ishay JS, Calton JG, Burnett JW (1983) Isolation of a lethal factor from venom of Vespa orientalis (oriental hornet) by affinity chromatography using cross-reactive monoclonal antibody.Toxicon 21(1),166-169

Scouten WH (1981) Affinity chromatography John Wiley and Sons, New York

O'Shannessy DJ, Quarles RH (1987) Labeling of the oligosaccharide moieties of immunoglobulins. J Immunol Methods 99:153-161

Simons K, Beam A (1969) Isolation and partial characterisation of the polypeptide chains in human ceruloplasmin. Biochim Biophys Acta 175:260-270

Solomon B, Koppel R, Schwartz F, Fleminger G (1989) 8th Int Symp on Affinity Chromatografy and Biological Recognition, Jerusalem p93

Ternynck T, Avrameas S (1969) Cross-linking of proteins with glutaraldehyde and its use for the preparation of immunoadsorbents. Immunochemistry 6:53-66

Ternynck T, Avrameas S (1971) Effect of electrolytes and of distilled water on antigen-antibody complexes. Biochem J 125:297

Turková J (1978) Affinity chromatography Elsevier

Turková J (1978) Affinity chromatography Elsevier
Turková J, Petkov L, Sajdok J, Kǎs J, Beneš MJ (1990) Carbohydrates as a tool for oriented immobilisation of antigens and antibodies. J Chromatogr 500:585-593
Vidal J, Godbillon G, Gadal P (1980) Recovery of active, highly purified phosphoenolpyruvate carboxylase from specific immunoadsorbent column. FEBS Letters 118:31-34
Waseem A, Salahuddin A (1983) Anomalous temperature dependence of the specific interaction of concavalin A with a multivalent ligand dextran. Biochim Biophys Acta 746:65-71
Werblin TP, Kim YT, Quagliata F, Siskind GW (1973) Control of antibody synthesis III. Changes in heterogenity of antibody affinity during the course of the immune response. Immunology 24(3):477-492
Wilchek M, Oka T, Topper YJ (1975) Structure of a soluble superactive insulin is revealed by the nature of the complex between cyanogen bromide - activated Sepharose and amines. Proc Natl Acad Sci 72(3):1055-1088

Chapter 4

Avidin-Biotin Immobilisation Systems

M. Wilchek and E. A. Bayer

Introduction

The avidin-biotin system has become an extremely useful tool for many different applications in the fields of biology, biochemistry and biotechnology. The unusual affinity (K_d 10^{-15} M) of avidin (or its bacterial relative, streptavidin) for the vitamin biotin forms the basis for using this system; the avidin-biotin complex serves as a general mediator which brings together a desired target within a biological system and an external probe. This approach has found broad application, mainly for localisation studies, medical diagnostics and recombinant genetics. The principles and scope of avidin-biotin technology have been extensively described and catalogued, and the reader is referred to the series of reviews which are available in the literature (Bayer and Wilchek 1978, 1980; Bayer et al. 1979; Wilchek and Bayer 1984, 1988, 1989a) as well as a book which has recently appeared (Wilchek and Bayer 1990a).

One area in which the use of the avidin-biotin system has notoriously lagged behind the other applications has been for the isolation and immobilisation of biologically active material. The problem in this context was purportedly (and perhaps ironically) the high affinity constant which was considered inappropriate for the final chromatographic step, namely, the release of the biotinylated molecule from the column. It was argued that simply another method for immobilisation (i.e. lacking inherent advantage) was not particularly required at the time.

Indeed, a large variety of methods has been described for immobilising macromolecules such as proteins and nucleic acids, and there are many applications for such immobilised systems. Most methods for immobilisation are chemical in nature; namely, an inert carrier is activated, and the molecule of choice is coupled usually via amino groups although other functional groups can be used for this purpose. The immobilised molecule is generally used for the isolation or sequestration of biologically active molecules (Wilchek et al. 1984).

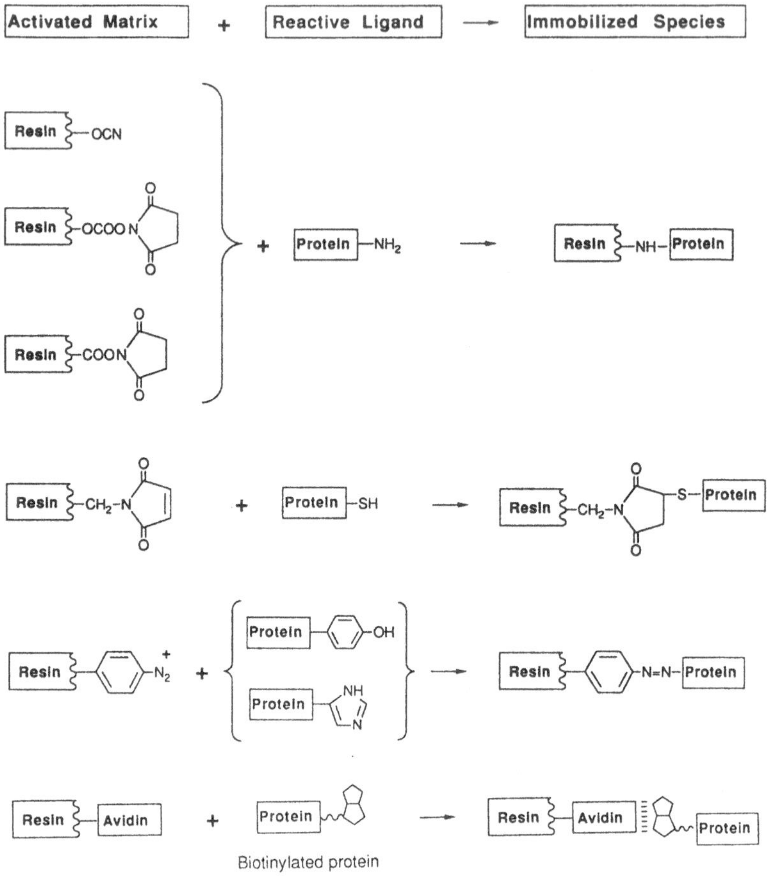

Fig. 4.1. Diagram illustrating avidin-biotin mediated immobilisation of proteins to resins versus various chemical methods for protein immobilisation.

In recent years, we have promoted the use of avidin-biotin technology for immobilisation studies (Wilchek and Bayer 1989b; Bayer and Wilchek 1990a,b). In this context, immobilised avidin can be considered a very stable "preactivated" carrier which binds biotin-containing molecules with exceptional affinity. For all practical purposes, the bonding between avidin and biotin is irreversible. Thus, an avidin-containing resin, which binds a biotinylated protein, can be considered to be similar to other conventionally activated matrices which react covalently with appropriate functional groups of a protein (See Fig. 4.1).

Since most biologists are uneasy with chemical reactions, immobilised avidin systems can be an excellent alternative; all that is required for coupling is to combine the biotinylated material and the avidin-containing resin. Finally, in view of the recent notion that biotin can be incorporated into proteins by recombinant DNA technology, the biotin moiety can be converted to comprise an "integral" part of a protein (Bayer

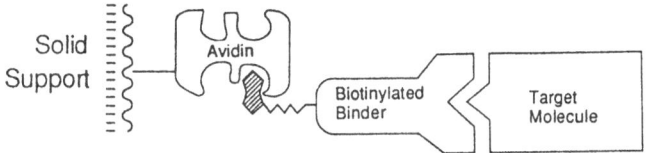

Fig. 4.2. Schematic description showing isolation of target molecule onto an avidin-biotin immobilised system. Note: the interaction between the immobilised avidin and the biotinylated binder is essentially a stable irreversible noncovalent complex, which permits subsequent elution of the target molecule in a purified state.

and Wilchek 1990c; Cronan 1990). Thus, a biologically active compound, which has been biotinylated by any method (Bayer and Wilchek 1990a), can be readily attached to the avidin column and used just like any other immobilised system (see Fig. 4.2).

Since the biotin moiety can be selectively incorporated into proteins to virtually any of their functional groups, the avidin-biotin system can actually be used as a substitute for all of the direct methods for covalent coupling with the added advantage of a built-in spacer. The only other possible cognate approach would be the use of a suitable hapten-antibody system (e.g. Kessler 1991), this being at the expense of reduced capacity, affinity and, hence, stability of the immobilising system.

Of course, in order to produce an avidin column, one must first immobilise the protein to a suitable carrier. In this review, we will describe the various types of avidin, we will then describe the different ways to prepare immobilised avidin (or its derivatives and analogues) and, finally, we will describe the different applications of the immobilised avidins.

Immobilisation of Avidins

Different Types of Avidin

Had this short review been written but a year ago, we would have described only the immobilisation to solid supports of the egg-white glycoprotein avidin and the bacterial protein streptavidin (Chaiet and Wolf 1964). In previous years, streptavidin was considered to be superior in many systems, due to the fact that avidin is highly charged (pI ~ 10.5) and bears carbohydrate residues; the "nonspecific" binding which often plagued the use of avidin-based probes was commonly attributed to these two properties of the glycoprotein. Thus, the net positive charge of avidin leads to undesirable electrostatic interactions and the presence of carbohydrate would render avidin interactive with lectins or other sugar-binding molecules in a biological system.

In order to counter the contribution of charge, the avidin molecule has been subjected to various types of chemical modification. This approach produced a large variety of avidin derivatives which boast neutral or slightly acidic pI values. A solution to the problem of the resident carbohydrate groups, however, was more elusive. Streptavidin, being a nonglycosylated neutral bacterial protein, was therefore used instead of the egg-white glycoprotein for most applications of avidin-biotin technology.

More recently, some latent structural impediments of the streptavidin molecule were described which, in retrospect, can account for the inexplicable "nonspecific" labeling

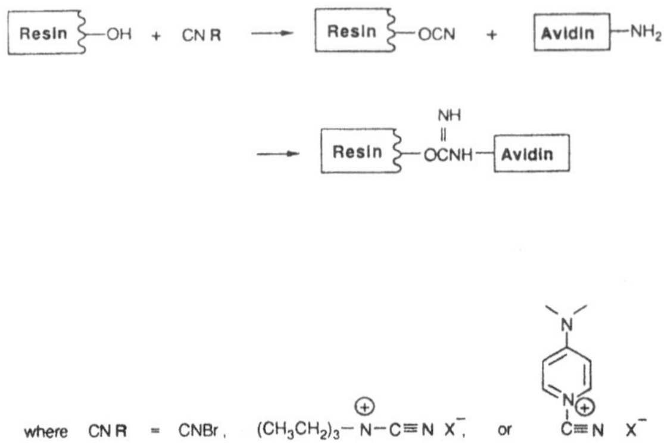

Fig. 4.3. Attachment of avidin via amino groups to resins by cyanylation.

which has at times (especially for intact cells and cell-derived material) accompanied streptavidin-based mediation (Alon *et al.* 1990). The cause of this unwanted label appears to be a function of the presence of an Arg-Tyr-Asp sequence in streptavidin, which mimics the concensus cell-recognition sequence, Arg-Gly-Asp (RGD), which characterises cell adhesion molecules like fibronectin (Ruoslahti 1990; Albelda and Buck 1990). Thus, streptavidin exhibits a defined interaction with the integrins and cells which bear these and similar receptors on their surface.

Avidin lacks an RGD-like sequence. Consequently, there is a trend to return to the egg-white component, but in modified form (there is also a financial consideration, since streptavidin is customarily much more expensive than avidin). Indeed, a method for converting the glycoprotein into a nonglycosylated derivative has been described (Hiller *et al.* 1987), and, during the last year, new and improved modified avidins have been introduced into the market. Among these include "Lite" Avidin (a sugar-less derivative of the native glycoprotein) and "NeutraLite" Avidin, which is a modified, nonglycosylated form of avidin, in which the net positive charge has been neutralised (both "Lite Avidin" and "NeutraLite Avidin" are trade names for nonglycosylated and neutralised nonglycosylated avidins, respectively, products prepared according to proprietary methods and available commercially through Belovo Chemicals, Industrial area 1, 6600 Bastogne, Belgium). Other derivatised forms of avidin still retain the carbohydrate chains and are thus inappropriate for use in many experimental systems. To date, it seems that streptavidin and other derivatives of egg-white avidin will eventually be substituted by NeutraLite Avidin, which appears to exhibit improved properties over all of the previously engineered avidins and streptavidins.

Immobilisation Techniques

All techniques which have been used for immobilisation of proteins are suitable for the immobilisation of avidin. Interestingly, none of the known immobilisation methods have caused a dramatic reduction in its binding capacity or apparent affinity for biotin.

The oldest, classical method for coupling to insoluble polysaccharides, the CNBr procedure (Fig. 4.3) was used in the initial works involving the avidin-biotin system,

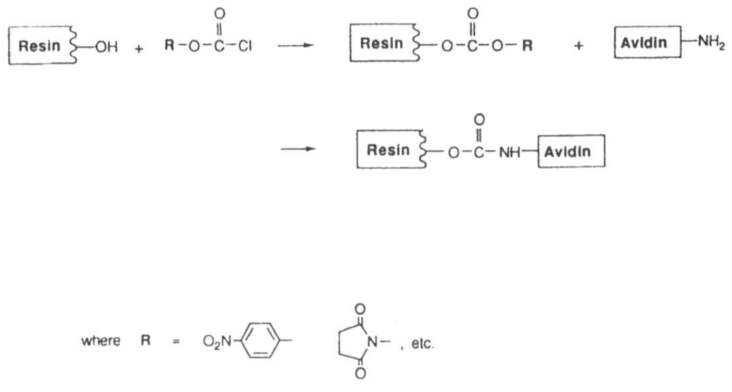

$$\boxed{\text{Resin}}\text{—OH} + \text{R—O—}\overset{\displaystyle O}{\overset{\|}{\text{C}}}\text{—Cl} \longrightarrow \boxed{\text{Resin}}\text{—O—}\overset{\displaystyle O}{\overset{\|}{\text{C}}}\text{—O—R} + \boxed{\text{Avidin}}\text{—NH}_2$$

$$\longrightarrow \boxed{\text{Resin}}\text{—O—}\overset{\displaystyle O}{\overset{\|}{\text{C}}}\text{—NH—}\boxed{\text{Avidin}}$$

where R = O_2N—⟨ ⟩—   ⟨N—⟩ , etc.

Fig. 4.4. Attachment of avidin via amino groups to chloroformate-activated resins.

particularly for the isolation or sequestration of native biotin-containing systems. Of the three different cyanylating reagents presented in the scheme, the dimethylaminopyridine derivative seems to be the safest and the best, since it is a stable crystaline compound (no fume hood is necessary) and high activation yields are consistently obtained (Kohn and Wilchek 1984).

In our hands, we have found that the cyanylating method gives satisfactory results, despite the current view that high levels of leakage often result, due to the formation of the unstable isourea bond. In the case of avidin, the reason for its unusual stability may reflect its tetrameric structure and the four biotin-binding sites. Stable crosslinking thus occurs upon interaction with the biotinylated compound. The stability of the avidin-biotin complex is induced by the cross-subunit interlocking by Trp-110 of the C-terminal tail of avidin, which, according to X-ray studies, fastens the biotin moiety within the binding site.

Using cyanylating reagents, the avidin remains highly charged. In order to further stablilise and decrease the number of charged groups, chloroformate-activated carriers have been used to immobilise avidin. For this purpose, hydroxy-containing polymers have been activated using chloroformates which contain good leaving groups (Wilchek and Miron 1982), as illustrated in Figure 4.4.

Binding of avidin using this method gives very high levels of stability (at least 40-fold that achieved by cyanylation-induced coupling), with respect to the bond between the carrier and the protein. As in the previous method, there is no reduction in the capacity of the immobilised avidin to bind biotinylated compounds, even though there is a reduction in the net charge of the system, in line with earlier evidence that lysines are not directly important for biotin binding.

A third method which has been successfully used for the attachment of avidin is via the Affigel system (BioRad). If an N-hydroxysuccinimide ester is prepared on a solid support as detailed earlier (Wilchek and Miron 1987), then avidin can be coupled via its lysine groups, yielding a very stable amide bond as shown in Figure 4.5. There are, theoretically, added advantages in this immobilisation approach, since the basicity of avidin is consequently reduced. Moreover, additional acidic groups are formed due to autohydrolysis of the active ester. Thus, the environment of avidin is much less alkaline which guards against nonspecific binding and aids in the subsequent elution step.

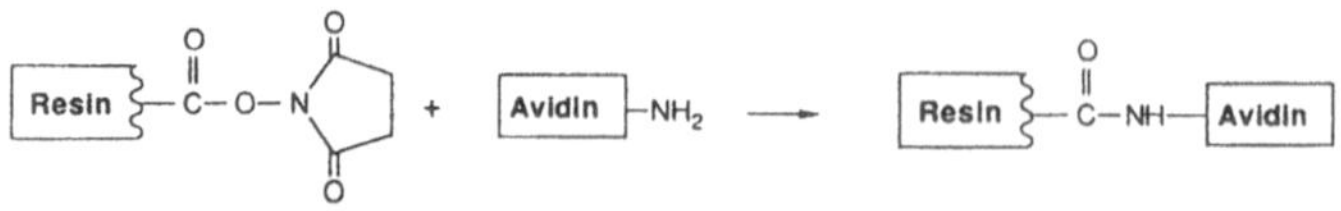

Fig. 4.5. Attachment of avidin via amino groups to N-hydroxysuccinimide-activated resins.

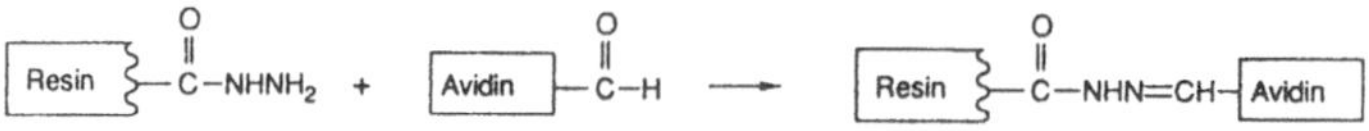

Fig. 4.6. Attachment of periodate-oxidised avidin to hydrazide-containing resins.

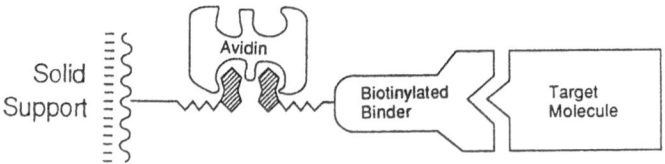

Fig. 4.7. Schematic description showing isolation of target molecule using an immobilisation system consisting of a biotinylated resin and a biotinylated binder, bridged by the free native avidin molecule, by virtue of its four biotin-binding sites.

Yet another approach for the immobilisation of avidin (in this case only native avidin and not streptavidin or nonglycosylated forms of egg-white avidin) is through the resident carbohydrate moieties. The sugars of avidin can be oxidised using periodate to yield aldehydes, and the oxidised glycoproteins can be coupled to a hydrazide-containing carrier according to Figure 4.6.

In our early attempts to use this method (Bayer and Wilchek 1980), a near-quantitative yield of coupling was achieved with a biotin-binding capacity of about 70% that of the native protein. These results are in accord with chemical modification studies on avidin using periodate (Green 1975). More recently, a similar procedure was described (O'Shannessy and Hoffman 1987), in which full binding capacity was reportedly retained. The reduced capacity in the earlier studies may be due to the oxidation of some of the tryptophans by periodate, and more controlled oxidation may be required.

One final method for immobilisation deserves special mention. In view of the tetrameric structure of the avidins, they can be readily immobilised to biotin-containing columns. In this case, there is no chemical reaction directly on the protein. Rather, the biotin moiety is chemically incorporated onto an appropriate resin, and the avidin serves as a bridge between the immobilised biotin and a biotinylated binder as shown in Figure 4.7.

The above immobilisation methods are the major methods by which avidin has been coupled to resins to date. In all cases, affinity and capacity are retained. If lower affinity is required (for example, for the desired release of biotin-containing compounds), avidin monomer columns may be used (Kohanski and Lane 1990). The preparation of these columns involves harsh treatment of the native immobilised tetramer, e.g. 3 M guanidinium isothiocyanate or 6 M guanidinium hydrochloride (each at pH 1.5), the action of which removes the uncoupled subunits. The remaining avidin monomer, attached covalently to the resin, lacks the adjacent subunit which would contribute Trp-110; the biotin is bound but not locked into the binding site, and, consequently, the affinity constant is dramatically reduced (ca., 10^{-7} M). This allows facile release of the biotinylated material, and, in many cases, free biotin can be used as an eluant.

Since the purpose of this review is to use avidin as an immobilisation technique, the avidin monomer columns will not be detailed further. Such columns would release both the target material together with the biotinylated binder, and their use would thus be restricted.

Applications

As an immobilising system, avidin-biotin technology was used mainly for the isolation of biologically active proteins (Wilchek and Bayer 1990b) and as a capture system for antigens and antibodies in immunoassay (Wilchek and Bayer 1990c) and for DNA (nucleic acid hybridisation technology).

In order to use this system, one of the interacting partners, the binder, has to be biotinylated to allow for its attachment to the avidin column for subsequent interaction with the desired target. During the years, different methods for biotinylating various functional groups of proteins, lipids, carbohydrates and nucleic acids have been described. In most cases, the biotinylated compound retained biological activity as long as the reaction was performed under mild and controlled conditions. For a review of the various biotinylating methods, the reader is referred to Wilchek and Bayer (1989a,1990d) and Bayer and Wilchek (1990d).

The first example in using such a system was given for the isolation of the insulin receptor using biotinylated insulin immobilised onto an avidin column (Finn et al. 1984). In this early study, the receptor was dissociated together with the biotinylated insulin, but in later studies conditions were established under which the receptor could be selectively released (Kohanski and Lane 1985); the immobilised insulin column could thus be reused.

Avidin columns were also used for the isolation of native biotin-containing peptides and proteins (Bodansky and Bodansky 1970; Lane et al. 1970; Berger and Wood 1975) and for the immobilisation of proteins (e.g. lectins and antibodies) in which biotin moieties were artificially incorporated (Bayer et al. 1976).

When an avidin-containing column is used for the purpose of isolation of the target molecule, the biotinylated compound can be applied directly to the avidin column without prior separation of nonbiotinylated material. In such cases, an inherent advantage in the system is that the avidin column will selectively remove the biotinylated compound. Another good practice in using this system for immobilisation is to guard against overloading the column with the biotinylated compound; this will have defined advantages in later stages of purification.

The above-described considerations are true for most biologically active proteins isolated via the avidin-biotin system. Some of the receptors which have been isolated using avidin-biotin technology are summarised in Table 4.1.

Table 4.1. Cell surface receptors isolated using avidin-biotin technology

Receptor	Reference
Immunoglobulin E	Lee and Conrad (1984), Nakajima and Delespesse (1986)
Estrogen	Redeuilh *et al.* (1985)
Glucocorticoid	Manz *et al.* (1983)
Gonadotropin releasing hormone	Hazum *et al.* (1986)
Insulin	Finn *et al.* (1984), Kohanski and Lane (1985)
Growth hormone	Haeuptle *et al.* (1983)
Opioid	Nakayama *et al.* (1986)
Cytomegaloviral proteins	Gretch *et al.* (1987)
G proteins	Kohnken and Hildebrandt (1989)
Angiotensin	Marie *et al.* (1990)
Endothelin	Wada *et al.* (1990)
Erythropoietin	Wognum *et al.* (1990)

Immobilised avidins can also be used for the adsorption of biotinylated forms of both single- and double-stranded DNA (Hultman *et al.* 1989, 1990). When using egg-white avidin for this purpose, one should remember that avidin is a strongly alkaline protein and reacts with nucleic acids via electrostatic interactions. Thus for all practical purposes, one should use streptavidin or neutralised forms of avidin (e.g. NeutraLite Avidin from Belovo Chemicals or ExtrAvidin from Sigma Chem. Co.). The immobilised biotinylated DNA can be subjected to different manipulative procedures, e.g. hybridisation, melting and interstrand disruption, without affecting the interaction between the avidin-biotin complex. Therefore, the immobilised avidin can be used for different purposes, particularly for the isolation of DNA-binding proteins, for the purification of RNA and single-stranded DNA, and for DNA sequencing (Hultman *et al.* 1991). It is interesting that many of the uses of streptavidin columns for DNA involve streptavidin immobilised to magnetic beads, which enables the rapid separation of the biotinylated DNA and interacting compounds. There is no doubt that the use of biotinylated DNA and immobilised avidins will flourish in the future.

Immobilised avidins can also be used for the isolation of synthetic peptides which have been biotinylated in the last step of synthesis using solid-phase methods (Bayer and Wilchek 1990a). Avidin carriers have also been used as an immobilising system for enzymes and hormones and their corresponding biological activities were in many cases conserved (Wilchek and Bayer 1989b).

Future Prospects

The use of avidin-biotin technology for isolation and immobilisation purposes has found extensive use in the research laboratory but has even greater potential. This system would in fact be very appropriate for industry, especially if the bond between avidin and biotin could be dissociated under conditions which would not destroy or inactivate either the biotinylated molecule or the immobilised avidin. If quantitative dissociation could be achieved, then the avidin column would be considered a "universal" column. The same column could be used repeatedly for different target material by adding the required biotinylated binder (Wilchek and Bayer 1989b).

References

Alon R, Bayer EA, Wilchek M (1990) Streptavidin contains an RYD sequence which mimics the RGD receptor domain of fibronectin. Biochem Biophys Res Commun 170:1236-1241

Albelda SM, Buck CA (1990) Integrins and other cell adhesion molecules. FASEB J. 4:2868-2880

Bayer EA, Wilchek M (1978) The avidin-biotin complex as a tool in molecular biology. Trends Biochem Sci 3:N237-N239

Bayer EA, Wilchek M (1980) The use of avidin-biotin complex in molecular biology. Methods Biochem Anal 26:1-45

Bayer EA, Wilchek M (1990a) The application of avidin-biotin technology for affinity based separations. J Chromatogr 510:3-11

Bayer EA, Wilchek M (1990b) Avidin column as a highly efficient and stable alternative for immobilisation of ligands for affinity chromatography. J Molec Recog 3:115-120

Bayer EA, Wilchek M (1990c) Biotin-binding proteins: Overview and prospects. Methods Enzymol 184:49-51

Bayer EA, Wilchek M (1990d) Protein biotinylation. Methods Enzymol 184:138-160

Bayer EA, Wilchek M, Skutelsky E (1976) Affinity cytochemistry: The localisation of lectin and antibody receptors on erythrocytes via the avidin-biotin complex. FEBS Lett 68:240-244

Bayer EA, Skutelsky E, Wilchek M (1979) The avidin-biotin complex in affinity cytochemistry. Methods Enzymol 62:308-315

Berger M, Wood HG (1975) Purification of the subunits of transcarboxylase by affinity chromatography on avidin-Sepharose. J Biol Chem 250:927-933

Bodansky A, Bodansky M (1970) Sepharose-avidin column for the biniding of biotin or biotin-containing peptides. Experientia 26:327

Chaiet L, Wolf FJ (1964) The properties of streptavidin - a biotin-binding protein produced by *Streptomyces*. Arch Biochem Biophys 106:1-5

Cronan JE Jr (1990) Biotination of proteins *in vivo*. J Biol Chem 265:10327-10333

Finn FM, Titus G, Horstman D, Hofmann K (1984) Avidin-biotin affinity chromatography: application to the isolation of human placental insulin receptor. Proc Natl Acad Sci USA 81:7328-7332

Green NM (1975) Avidin. Adv Protein Chem 29:85-133

Gretch DR, Suter M, Stinski MF (1987) The use of biotinylated monoclonal antibody and streptavidin affinity chromatography to isolate herpesvirus hydrophobic proteins or glycoproteins. Anal Biochem 163:270-277

Hazum E, Schvartz I, Waksman Y, Keinan D (1986) Solubilisation and purification of rat pituitary gonadotropin-releasing hormone receptor. J Biol Chem 261:13043-13048

Haeuptle M-T, Aubert ML, Djiane J, Kraehenbuhl J-P (1983) Binding sites for lactogenic and somatogenic hormones from rabbit mammary gland and liver. J Biol Chem 258,305-314

Hiller Y, Gershoni JM, Bayer EA, Wilchek M (1987) Biotin binding to avidin: oligosaccharide side chain not required for ligand association. Biochem J 248:167-171

Hultman T, Stahl S, Hornes E, Uhlen M (1989) Direct solid phase sequencing of genomic and plasmid DNA using magnetic beads as solid support. Nucleic Acids Res 17:4937-4946

Hultman T, Stahl S, Hornes E, Uhlen M (1990) Solid phase in vitro mutagenesis using plasmid DNA template. Nucleic Acids Res 18:5107-5112

Hultman T, Bergh S, Moks T, Uhlen M (1991) Bidirectional solid phase sequencing of in vitro amplified plasmid DNA. BioTechniques 10:84-93

Kessler C (1991) The digoxigenin:anti-digoxigenen (DIG) technology - a survey on the concept and realisation of a novel bioanalytical indicator system. Molec Cellul Probes 5:161-205

Kohanski RA, Lane MD (1985) Homogeneous functional insulin receptor from 3T3-L1 adipocytes. J Biol Chem 260:5014-5025

Kohanski RA, Lane MD (1990) Monovalent avidin affinity column. Methods Enzymol 184:194-200

Kohn J, Wilchek M (1984) The use of cyanogen bromide and other novel cyanylating agents for the activation of polysaccharide resins. Appl Biochem Biotechnol 9:285-305

Kohnken RE, Hildebrandt JD (1989) G Protein subunit interactions, studies with biotinylated G protein subunits. J Biol Chem 264:20688-20696

Lane MD, Edwards J, Stoll E, Moss J (1970) Tricarboxylic acid activator-induced changes at the active site of acetyl-CoA carboxylase. Vitam Horm 28:345-363

Lee WT, Conrad DH (1984) The murine lymphocyte receptor for IgE. J Exp Med 159:1790-1795

Manz B, Heubner A, Kohler I, Grill H-J, Pollow K (1983) Synthesis of biotin-labelled dexamethasone derivatives. Eur J Biochem 131:333-338

Marie J, Seyer R, Lombard C, Desarnaud F, Aumelas A, Jard S, Bonnafous J-C (1990) Affinity chromatography purification of angiotensin II receptor using photoactivable biotinylated probes. Biochemistry 29:8943-8950

O'Shannessy DJ, Hoffman WL (1987) Site-directed immobilisation of glycoproteins on hydrazide-containing solid supports. Biotechnol Appl Biochem 9:488-496

Nakajima T, Delespesse G (1986) IgE receptors on human lymphocytes. Eur J Immunol 16, 809-814

Nakayama H, Shikano H, Aoyama T, Amano T, Kanaoka Y (1986) Affinity purification of the opioid receptor in NG108-15 cells using an avidin-biotin system with a novel elution method. FEBS Lett 208:278-282

Redeuilh G, Secco C, Baulieu EE (1985) The use of the biotinyl estradiol-avidin system for the purification of "nontransformed" estrogen receptor by biohormonal affinity chromatography. J Biol Chem 260:3996-4002

Ruoslahti E (1990) Versatile mechanisms of cell adhesion. The Harvey Lectures, Series 84, 1-17

Wada K, Tabuchi H, Ohba R, Satoh M, Tachibana Y, Akiyama N, Hiraoka O, Asakura A, Miyamoto C, Furuichi Y (1990) Purification of an endothelin receoptor from human placenta. Biochem Biophys Res Commun 167:251-257

Wilchek M, Bayer EA (1984) The avidin-biotin complex in immunology. Immunol Today 5:39-43

Wilchek M, Bayer EA (1988) The avidin-biotin complex in bioanalytical applications. Anal Biochem 171:1-32

Wilchek M, Bayer EA (1989a) Avidin-biotin technology ten years on: has it lived up to its expectations? Trends Biochem Sci 14:408-412

Wilchek M, Bayer EA (1989b) A universal affinity column using avidin-biotin technology. In: Hutchens TW (ed) Protein recognition of immobilised ligands. Alan R Liss, Inc New York, pp 83-90

Wilchek M, Bayer EA (1990a) Avidin-biotin technology (Methods Enzymol, vol 184) Academic Press, Orlando

Wilchek M, Bayer EA (1990b) Isolation of biologically active compounds: a universal approach. Methods Enzymol 184:243-244

Wilchek M, Bayer EA (1990c) Avidin-biotin mediated immunoassays: Overview. Methods Enzymol 184:467-469

Wilchek M, Bayer EA (1990d) Biotin-containing reagents. Methods Enzymol 184:123-138

Wilchek M, Miron T (1982) Immobilisation of enzymes and affinity ligands onto agarose via stable and uncharged carbamate linkages. Biochem Internat 4:629-635

Wilchek M, Miron T (1987) Limitations of N-hydroxysuccinimide esters in affinity chromatography and protein immobilisation. Biochemistry 26:2155-2161

Wognum AW, Lansdorp PM, Humphries RK, Krystal G (1990) Detection and isolation of the erythropoietin receptor using biotinylated erythropoietin. Blood 76:697-705

Chapter 5

Development of Affinity Membranes

P. W. Feldhoff

Introduction

In the development of affinity membrane technology, several areas of expertise come together and each has a basis of knowledge that contributes to the goal of affinity membrane separation (Fig. 5.1).

In the area of membrane science there is the choice of polymer, membrane configuration and pore size. Once the matrix considerations have been decided, chemistry and biochemistry expertise are necessary in the choice of membrane activation chemistry, attachment of ligand and biological applications, i.e. ligand/ligate interactions.

Basically two chemical processes are required to chemically immobilise a ligand onto an inert matrix. First, a reactive group of defined chemical function must be introduced into the inert matrix yielding an activated membrane. Second, the ligand either binds to the reactive group or displaces the reactive group, in either case the ligand is chemically immobilised on the matrix. In the case of affinity membranes the ligand has biospecificity. The choice of ligand attached is based on biospecific interaction with the ligate which is the molecule one wishes to pull out of solution. The ability of the ligand and ligate to interact will depend on their biospecificity and the complexity of the feed solution.

After initial development of the affinity membrane the chemical and process engineers must then develop reactor design and carry out process studies.

Matrix

Considering the choice of matrix, a microporous membrane with large pores, high porosity and thick walls would be ideal in providing high surface area while still permitting permeation of large molecules. Since an extended surface is needed to provide binding sites for ligands which can be accessed in flow conditions, a

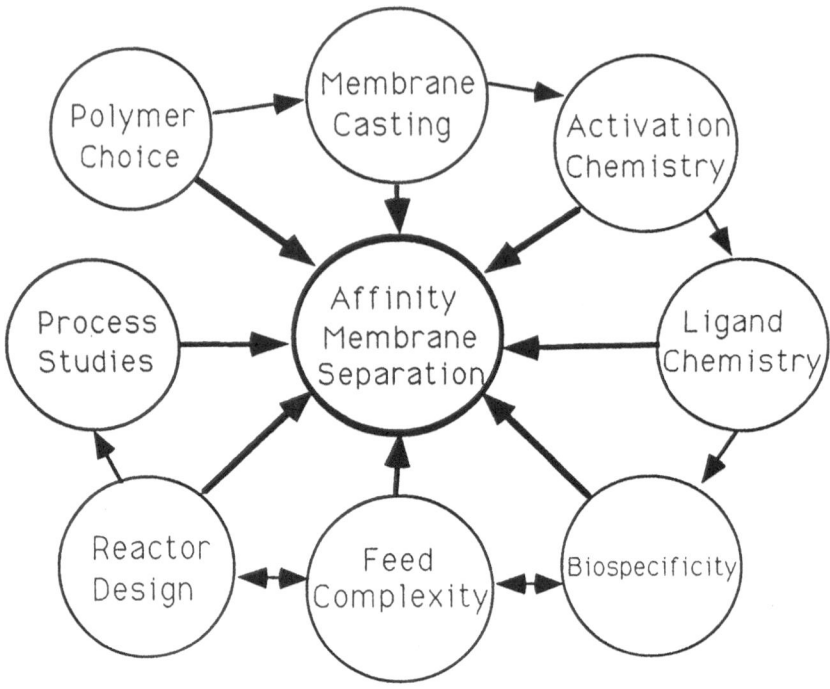

Fig. 5.1. Scheme of Affinity Membrane Development.

microporous membrane in a hollow fiber configuration operated under cross-flow conditions would be optimal for affinity membrane separations. The membrane must have sufficient functionality for ligand attachment, however, in some cases limited functionality can be amplified. In order to have the adsorptive specificity desired in affinity interactions, nonspecific binding should be kept at a minimum. The membrane itself should be uncharged and charge should not be introduced by the activation chemistry, thus avoiding nonspecific ionic interactions. Additionally the membrane should be hydrophilic in nature to avoid non-specific hydrophobic adsorption. Chemical reactions that would cross-link the matrix surface or shrink pore size should be avoided, in order to maintain high flow and elution conditions. The ideal matrix requirements are summarised in Table 5.1.

Table 5.1. Ideal matrix requirements

Large internal surface area, i.e. microporous
Adequate concentration of groups for activation
Minimal non-specific binding
    Uncharged to limit ionic interactions
    Hydrophilic to limit hydrophobic binding
Stable to solvents and chemical reagents
Mechanically and morphologically stable

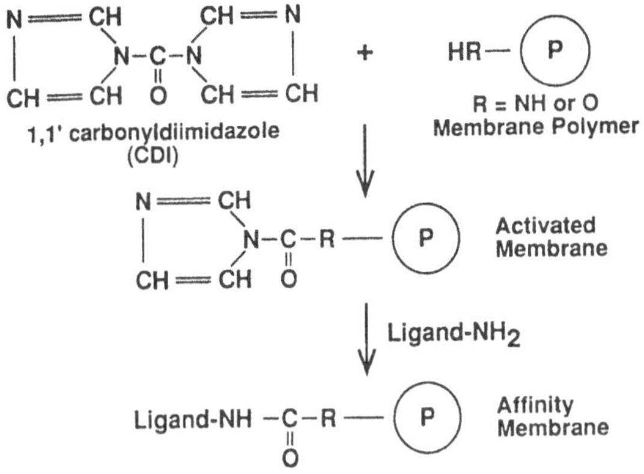

N══CH   CH══N
| \   / |
| N-C-N | +   HR—(P)
CH══CH   ‖   CH══CH
         O   R = NH or O
1,1' carbonyldiimidazole   Membrane Polymer
        (CDI)

                    ↓

N══CH
| \
| N-C-R —(P)   Activated
CH══CH   ‖   Membrane
         O

                    ↓ Ligand-NH$_2$

Ligand-NH —C-R —(P)   Affinity
          ‖   Membrane
          O

Fig. 5.2. Carbonyldiimidazole activation chemistry.

Activation Chemistries

Once the matrix is chosen, it is activated by the introduction of an electrophilic group into the matrix to make the functional groups of the matrix more reactive toward the nucleophiles of the ligand. The chemical activation occurs throughout the microporous membrane structure affording a large activated internal surface area. I have utilised three different activation chemistries to introduce reactive groups into polyamide and cellulosic membranes: carbonyldiimidazole (CDI) (Bethell et al. 1979), 2-fluoro-1-methyl-pyridinium (FMP) (Ngo 1986) and cyanuric chloride (CyCl$_3$)(Kay and Crook 1967). The three reagents were found to yield efficient membrane activation, to form stable covalent linkage between the ligand and the matrix, and do not introduce any charge or hydrophobic groups to the matrix. Each leads to the formation of a different type of covalent linkage between the ligand and the matrix (Feldhoff 1992).

The CDI reacts with the membrane to give an imidazolyl intermediate. The imidazole ring is then displaced by the amine function of a protein ligand (Fig. 5.2). CDI activation yields a covalent N-alkyl carbamate (urethane) linkage between the amine of the protein ligand and a cellulosic polymer and, an urea linkage between the ligand and the polyamide polymer.

CDI is water labile, therefore, activation is carried out in a non-aqueous solvent. We have utilised dioxane, acetone, acetonitrile, cyclohexane, freon and dimethyl sulfoxide as solvents. All lead to satisfactory activation. Due to cost and environmental considerations acetone is our solvent of choice. To maintain pore integrity, the membrane is gradually dehydrated prior to CDI activation. CDI is reactive, 15 min is sufficient for activation; nonionic, adding no charge to the polymer; and forms a stable covalent bond between the matrix and the ligand. The addition of ligand to the activated intermediate is carried out under aqueous conditions, 0.1 M bicarbonate buffer at pH 8 - pH 10. We routinely use pH 8.5. Though the activated intermediate is water labile, aqueous conditions are used for attachment of ligand to avoid cross-linking, denaturing or hydrolyzing the proteinaceous ligand. The CDI intermediate is so reactive that it provides fast coupling of ligand even in an aqueous environment. Results with 1,6-diaminohexane showed equivalent binding from acetone or 0.1 M bicarbonate

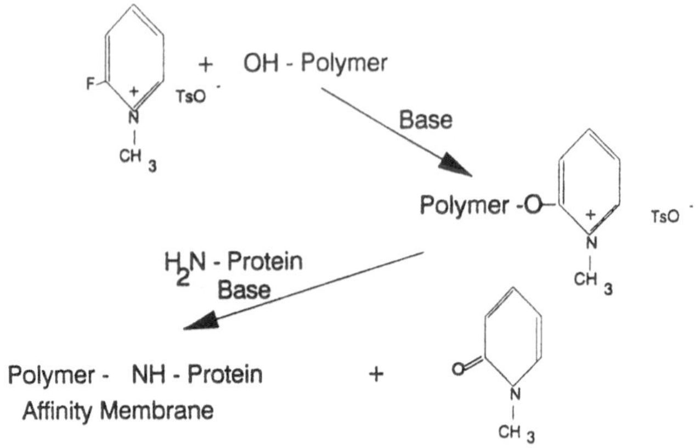

Fig. 5.3. 2-fluoro-1-methyl pyridinium toluene-4-sulfonate activation chemistry.

buffer, pH 8.5. Binding of this small ligand was complete in 15 min. Once the ligand is covalently bound to the membrane, it is washed with 0.1 M sodium acetate buffer to remove any unbound ligand.

FMP reacts with a cellulosic material to form an activated polymer intermediate (Fig. 5.3). The ring structure of the FMP acts as a leaving group and is displaced by the N-nucleophile of the peptide or protein to yield a stable amide bond between the carbon of the matrix and the amine of the ligand.

Thiols of the ligand will also bind to the FMP-activated polymer. The activation is carried out in organic solvent in the presence of triethylamine. Addition of ligand is routinely from basic aqueous conditions. We utilise a pH of 8-8.5, the range is pH 6.5 - pH 9.5. FMP is reactive and forms a stable covalent nonionic link between ligand and matrix.

In the case of CyCl$_3$, a chlorine is displaced in the activation of the polymer and the CyCl$_2$ is attached to the matrix. A second chlorine then acts as a leaving group when the ligand is attached to the activated matrix. The third chlorine is displaced with ethanolamine to eliminate unwanted reactions, leaving the inert ring structure of the CyCl as part of the matrix-ligand linkage (Fig. 5.4).

The covalent nature of the three types of linkages illustrated leads to the production of affinity membranes which retain capacity and effective long term operation. Preliminary results with these procedures have provided an affinity hollow fiber matrix that has been shown to maintain its membrane properties and to have a ligand concentration comparable to that found in equivalent gels. Choice of activation chemistry will depend on the reactivity, stability and local ionisation atmosphere of the particular membrane/ligand. Once the activation chemistry has been selected, the optimal conditions for concentration of activating agent, time of activation and temperature can be determined. The aim is an activation level which will yield the best protein capacity and greatest purification factor. This may not be the highest degree of activation. A membrane that is too highly activated may lead to multi-point attachment

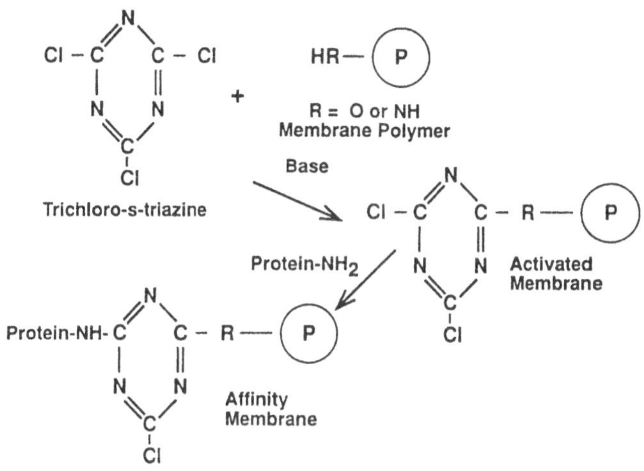

Fig. 5.4. Cyanuric chloride activation chemistry.

of ligand. Multi-point attachment would not be a problem when attaching a small peptide ligand but there is a potential problem in a protein which in addition to the N-terminus has, on the average, 7% lysine residues (Dayhoff and Hunt 1972) as potential attachment sites. If multi-point attachment does occur it may lead to improper orientation or altered conformation of the ligand and interfere with ligate binding. This would decrease function. Therefore, one wants a density and orientation of ligand which will allow good ligand/ligate interaction. The density of ligand attachment to the matrix will depend on the concentration of activated functional groups on the matrix, the size of ligand, the concentration of ligand offered and the reaction efficiency of the membrane intermediate with the ligand.

Table 5.2. Ligand/ligate interactions

Type	Ligand	Ligate
Class	Lectins	Sugars
	Protein A	Immunoglobulins
	Protein G	Immunoglobulins
	Heparin	Coagulation Factors
		Lipoprotein
	Polymyxin B	Endotoxin
	Benzamidine	Serine Proteases
Specific	Antibody	Antigen
	Antigen or Hapten	Antibody
	Enzyme	Inhibitor
	Artificial Substrate	Enzyme
	Inhibitor	Enzyme
	Co-factor	Enzyme
	Hormone	Receptor

Ligand/Ligate Interactions

Once the affinity membrane has been produced, ligate capture can be effected. The ligate solution should contain 0.1 M - 0.5 M salt to avoid nonspecific protein-protein interaction. The ligate solution can be presented to the affinity matrix in a single pass or the solution can be recirculated. Since the membrane-ligand/ligate interactions occur by convective rather than diffusive mass transfer, the efficiency of the loading process is limited only by the kinetics of the ligand/ligate interaction. The affinity membrane will allow for high separation efficiency with high throughput but sufficient residence time must be allowed to afford ligate capture.

The ligand/ligate interaction must be biospecific and reversible. The biospecificity can be general by class of compound or more limited and specific. Some examples of biospecific recognition are given in Table 5.2. As is indicated, the ligand is not required to be a protein or peptide but is limited only to the extent that the compound must have a nucleophilic group capable of displacing the leaving group of the activated membrane intermediate.

Once the ligand sites are saturated or the ligate solution depleted, the affinity membrane is washed to remove unwanted species and the ligate eluted. The elution method depends on the ligand/ligate interaction. Since the ligand/ligate interaction is based on sterospecific recognition, the ligate can be eluted by altering the conformation. This is accomplished by increasing the ionic strength or lowering the pH. Alternatively, the ligate can be displaced from the affinity membane by competition with an excess of a ligate analogue. Elution should be under conditions that do not denature or decrease the biological activity of the ligate. For example, if pH is used to elute the ligate, the pI of the ligate should be avoided and the eluant should be neutralised immediately. The cycle is completed and prepared for reuse by equilibration of the affinity membrane with the starting ionic strength and pH.

Utilising an affinity membrane one can isolate a protein from a high concentration of contaminating proteins, or effectively concentrate the ligate from a solution containing a low concentration of the protein of interest. Alternatively, affinity separation can be used to bind all components of the mixture except the molecule of interest. An application of the latter would be the removal of a low level contaminate of a protein preparation, for example removal of transferrin and alpha$_1$-antiproteinase, which are glycoproteins, from an albumin preparation using a concanavalin A-membrane.

Spacers and the Chemistry Involved

Depending on the ligand/ligate involved it may be desirable to extend the ligand from the membrane. The use of a spacer is critical for a low molecular weight ligand reacting with a large ligate, i.e. a small antigen reacting with an antibody, and in low affinity ligand/ligate interactions. The ligand can be attached to the membrane through a spacer which is generally a hydrocarbon chain of two to ten carbons with a reactive terminal group. The spacer is covalently attached to the membrane by one of the chemistries described leaving a free amino or carboxyl group. The ligand is then covalently linked to the spacer. Commonly used spacers and the possible spacer/ligand attachment chemistry are described in Table 5.3.

Table 5.3. Common spacers

Spacer	Chemistry
1,6 Diaminohexane	Reductive amination with an aldehyde or
	Condensation with a carboxyl
6 Aminohexanoic acid	Condensation with an amine
Adipic acid dihydrazide	Schiff base with aldehyde reduced with cyanoborohydride

The condensation reaction with 1,6 diaminohexane or 6-aminohexanoic acid and the ligand is most conveniently carried out with a water soluble carbodiimide, such as 1-ethyl-3-(3-dimethylaminopropyl) carbodiimide (EDAC). The choice of spacer is determined by the biospecificity of the ligand/ligate interaction, for example, if there is a requirement for the availability of the ligand carboxyl or amino terminal group for ligate binding. Unless there is a specific requirement for the amino terminal to be free, 6-aminohexanoic acid would be preferable as a spacer. In the latter case, the spacer is covalently linked to the membrane and the carboxyl group of the spacer activated with EDAC, excess EDAC is washed out and the ligand attached. The ligand is not present in solution with the EDAC during the condensation reaction, preventing activation of the ligand's carboxyl groups and possible cross-linking of the ligand.

Adipic acid dihydrazide is conveniently used as a spacer in the attachment of glycoproteins to a cellulosic membrane through the carbohydrate moiety of the glycoprotein. The adipic acid dihydrazide is attached to the membrane, the hydrazine reacts with the sugar aldehyde to form a Schiff base which is then reduced with sodium cyanoborohydride. Attachment of glycoproteins through the carbohydrate has been used successfully with IgG, providing IgG attachment without interfering with the antigen recognition site.

Spacers can be used to allow for better ligand/ligate interaction resulting in an increased ligate yield and a shortened residence time. Additionally, the membrane/spacer will yield alternate functional groups on the membrane allowing alternate attachment chemistries, so one is not limited to the chemistry of the base polymer.

Amplification of Matrix Sites

Another base polymer limitation, which may occur, is that enough sites may not exist on the membrane for chemical activation and optimal ligand attachment. Working with polyamide, we have developed methods to expand available sites for ligand attachment. The goal was to effectively increase membrane affinity capacity. Two experimental procedures were investigated. In the first, polyamide was modified by limited acid hydrolysis. This method was found to yield a 5-fold increase in the available sites for activation. However, though the integrity of the membrane was maintained under the mild conditions used, this is not an ideal manner to process a membrane. An alternate procedure, used to introduce more functional groups into the polyamide membrane, was to covalently link a neutral hydrophilic carbohydrate layer to the matrix (Klein and Feldhoff 1991). The amplified membrane is then activated for ligand attachment. This procedure not only yields multiple hydroxyl sites for activation, but also adds a hydrophilic layer to the polymer. An example of functional group amplification is depicted in Figure 5.5.

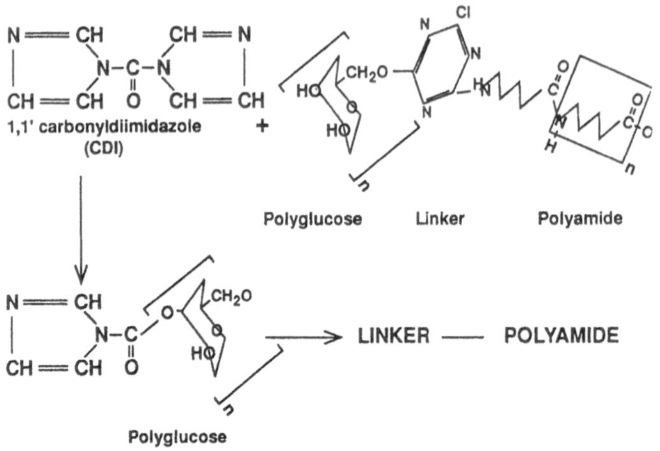

N=CH   CH=N
N-C-N
CH=CH O   CH=CH
1,1' carbonyldiimidazole   +
(CDI)

Polyglucose   Linker   Polyamide

N=CH
N-C-
CH=CH O

Polyglucose

LINKER —— POLYAMIDE

Fig. 5.5. Functional group amplification of the polyamide matrix. Polyglucose is covalently bound to the cyanuric chloride activated membrane. The amplified polyamide matrix is subsequently activated with CDI for ligand attachment.

Conclusion

Affinity membranes have advantages over the conventional affinity gels (Table 5.4).

Table 5.4. Advantages of Affinity Membranes Over Affinity Resin Columns

Suitable for large scale separations
High separation efficiency and volume throughput
Convective, high speed separations
Can handle dispersions
May condense purification scheme by eliminating initial isolations steps
Permits cross-flow operations
Low pressure drops

An affinity membrane is equivalent to an affinity gel column of very large diameter and small bed height with the advantage that the convective high speed separations of membranes overcome the diffusive mass transfer constraints of gels. Additionally, membranes allow high volume throughput, are resilient, and offer a reduced cost in scale-up operations when compared to gel systems. Affinity membranes would be especially useful in process scale applications such as harvesting recombinant DNA or cell culture products.

References

Bethell GS, Ayers JS, Hancock WS, Heam MTW (1979) A novel method of activation of cross-linked agaroses with 1,1'-carbonyldiimidazole which gives a matrix for affinity chromatography devoid of additional charged groups. J Biol Chem 254: 2572-2574

Dayhoff MO, Hunt LT (1972) Composition of proteins. In: Dayhoff MO (ed) Atlas of protein sequence and structure, vol 5. Natl Biomed Res Found, Silver Spring, MD, p D355

Feldhoff PW (1992) Comparison of coupling procedures for development of affinity membranes: optimisation of the CDI method. In: Angeletti RH (ed) Techniques in protein chemistry vol 3. Academic Press Inc, Orlando, Fl, P151-160

Kay G, Crook EM (1967) Coupling of enzymes to cellulose using chloro-s-triazines. Nature 216: 514-515

Klein E, Feldhoff PW (1991) US Patent serial 477512. Affinity separation with activated polyamide mecioporous membranes

Ngo TT (1986) Facile activation of sepharose hydroxyl groups by 2-fluoro-1-methylpyridinium toluene-4-sulfonate: preparation of affinity and covalent chromatography matrices. Bio-Technology 4: 134-137

Chapter 6

Crystalline Protein Layers as Isoporous Molecular Sieves and Immobilisation and Affinity Matrices

M. Sára, S. Küpcü, C. Weiner, S. Weigert and U. B. Sleytr

Introduction

During the last 30 years membrane technology has developed into a topic of major practical and theoretical importance. Research in membrane technology involves the disciplines of physics and chemistry on the one hand and biochemical and process engineering on the other hand. A major breakthrough in the history of membrane technology was the development of asymmetric membranes where a thin selective layer of the order of 0.1 μm to 1 μm is supported by a highly porous sublayer of the same material (for reviews see Lonsdale 1981; Strathmann 1982). The symmetric membranes which consisted of an approximately 100 μm thick selective layer produced before were never really applied in industrial and laboratory processes. Another breakthrough was the development of composite membranes with an asymmetric structure where a top layer is supported by a porous sublayer. In this case the two layers consist of different polymeric materials. The advantage of composite membranes is that each layer can be optimised independently to obtain optimal membrane performance with respect to selectivity, permeation rate and chemical stability. Nowadays membrane processes are classified into microfiltration, ultrafiltration and hyperfiltration. Microporous membranes with a pore size from > 50 nm to 2 μm are applied in microfiltration processes and mesoporous membranes with a pore size from 2 nm to 50 nm are used for ultrafiltration, whereas in hyperfiltration non porous solution diffusion membranes are used.

Ultrafiltration is used to retain macromolecules from a solution. The rejection characteristics of ultrafiltration membranes are theoretically determined by the size and shape of the solutes relative to the pore size in the permselective layer (Blatt 1976; Mulder 1991). Ultrafiltration membranes are produced from various polymers such as cellulose derivatives, polysulfone or ionic polymers by the phase inversion process.

Although the pores with a size of 2 nm to 50 nm are in the resolution range of transmission electron microscopes it has not been possible to visualise them by applying this technique so far. The major reason can be seen in the amorphous structure of the active filtration layer which should be rather considered as a heteroporous network of polymer chains than as a closed layer with defined openings. Although most recently the pores exposed on the surface of an ultrafiltration membrane could be visualised by atomic-force microscopy (Dietz *et al.* 1991) the average pore size is usually derived from the specific rejection characteristics of the membranes.

In this chapter we will describe the unique molecular sieving properties of isoporous ultrafiltration membranes produced from crystalline bacterial cell surface layers. Further we will discuss the broad spectrum of chemical modification reactions which can be applied for changing the nominal molecular weight cut off and the physicochemical properties of the crystalline lattices. In the last section the use of the bacterial cell surface layers as immobilisation and affinity matrix is described.

Crystalline Bacterial Cell Surface Layers

General Properties

Crystalline bacterial cell surface layers (S-layers) represent the outermost cell wall component in many eubacteria and are a fairly common feature of the archaebacterial cell wall (for compilation see Messner and Sleytr 1992; Sleytr and Messner 1983; Sleytr *et al.* 1988). S-layers are composed of an identical protein or glycoprotein species showing apparent molecular weights of 40 000 to 220 000. S-layer lattices can show oblique (p2), square (p4) or hexagonal (p3, p6) symmetry (for reviews see Beveridge 1981; Baumeister and Engelhardt 1987; Hovmöller *et al.* 1988; Kandler and König 1985; Koval 1988; Messner and Sleytr 1992; Sleytr and Messner 1983, 1988; Smit 1987). One morphological unit is composed of two (p2), three (p3), four (p4) or six (p6) identical subunits. The centre to centre spacing of the morphological units is a strain specific feature and lies between 3 nm and 32 nm (for a more detailed description see Pum *et al.* Chapter 10 this volume).

In the case of gram-positive eubacteria the subunits are linked to each other and to the underlying peptidoglycan-containing layer by noncovalent interactions including hydrogen bonds, direct electrostatic interactions between free amino and carboxyl groups and hydrophobic bonds. Therefore, isolation of S-layer subunits from cell wall fragments is possible by changing the pH or by applying high concentrations of hydrogen bonds breaking agents. The peptidoglycan-containing sacculi can be simply removed by a single centrifugation step whereas the solubilised S-layer subunits remain in the clear supernatant. S-layer subunits frequently reassemble into regularly structured lattices identical to those observed on intact cells if the disrupting agent is slowly removed, e.g. during dialysis. Usually, the formed S-layer self-assembly products are flat sheets or open-ended cylinders which can attain a size of up to 10 μm (for review see Sleytr and Messner 1989).

S-layer proteins have a high level of acidic amino acids. Thus, the pI of various S-layer proteins is in the range of 3.5 to 5.5. Only a low amount of arginine or histidine have been detected but all S-layer proteins investigated so far revealed a high amount (up to 10%) of lysine. As most bacterial exoproteins, S-layer proteins lack sulfur-containing amino acids.

Permeability Properties of S-layers

High-resolution electron microscopical studies on the mass distribution in S-layers of different organisms have indicated pores with a size from 2 nm to 8 nm (Aebi *et al.* 1973; Beveridge 1979; Burley and Murray 1983; Deatherage *et al.* 1983; Dickson *et al.* 1986; Engelhardt *et al.* 1986; Lepault *et al.* 1986; Sjögren *et al.* 1985; Sleytr and Messner 1983; Stewart and Beveridge 1980; Stewart *et al.* 1980; Stewart and Murray 1982). As expected for two-dimensional crystalline lattices composed of identical subunits the pores showed identical size and morphology. In many S-layers two or even more distinct classes of pore can be observed (Hovmöller *et al.* 1988).

Although electron microscopical studies provide valuable information on the mass distribution in S-layer lattices it is not yet possible to predict the actual molecular-sieving properties of the crystalline arrays from these data. In the course of specimen preparation many different structural changes including denaturation and collapse phenomena may occur in the delicate S-layer protein network. Even after chemical fixation of the S-layer protein with glutaraldehyde native domains still capable of undergoing conformational changes upon incubation in solutions of different pH or ionic strength can be maintained (Sára unpublished observation). For example, permeability studies on glutaraldeyhde-treated S-layers from thermophilic Bacillaceae revealed that the pores became significantly smaller upon lowering the pH from 7 to 3.5. Such alterations can be considered to occur in the course of the common negative-staining procedure as used for high-resolution electron microscopical studies. Furthermore, with negatively-stained preparations there is still the uncertainty in defining the boundaries between the heavy metal stain which penetrates the protein free areas and the protein itself (Engelhardt 1988). Usually with the digital image reconstruction procedures used for determining the mass distribution in S-layer lattices, the steepest gradient in contrast between protein and stain is chosen as threshold within the limits of a contiguous network. Thus it can be concluded that electron microscopy can provide information on only a "static" morphological state of the pores in the S-layer lattice. To obtain information on the "dynamic" and physicochemical properties of the pores including the influence of charged and hydrophobic groups on the molecular-sieving properties permeability studies have been carried out. These experiments were performed with native and glutaraldehyde-treated S-layer vesicles using structurally well characterised solutes as test molecules (Sára *et al.* 1987a). By applying the "space technique" (Scherrer and Gerhardt 1971), S-layers from thermophilic Bacillaceae showed sharp exclusion limits between molecular weights of 30 000 and 45 0000, suggesting a limiting pore diameter of 4 nm to 5 nm (Sára and Sleytr 1987a). Carbonic anhydrase with a molecular weight of 30 000 and a molecular size of 4.1 nm x 4.1 nm x 4.7 nm could still pass through the pores whereas ovalbumin with a molecular weight of 43 000 was rejected to at least 90%. No difference in the rejection characteristics could be observed between native and glutaraldehyde-treated S-layer vesicles.

Physicochemical Properties of S-layers

When considering the physicochemical properties of the S-layers it is necessary to distinguish between the outer S-layer surface exposed to the ambient environment and the inner S-layer surface facing the underlying cell envelope components such as the peptidoglycan-containing layer in the gram-positive eubacteria. Labelling experiments with strongly positively charged topographical marker molecules, such as polycationised ferritin (pI>11) (Danon *et al.* 1972), revealed that the outer face of the S-layer lattice from several thermophilic and mesophilic Bacillaceae was not negatively charged (Sára and Sleytr 1987b; Sára *et al.* 1990). Chemical modification reactions applied to different S-layer proteins confirmed that amino and carboxyl groups present on the outer S-layer face neutralise themselves by forming direct electrostatic interactions. By contrast, the inner face of the S-layer lattice showed an excess of free carboxyl groups (Pum *et al.* 1989; Sára and Sleytr 1987b). Studies on the S-layers from thermophilic Bacillaceae also demonstrated the presence of free amino and carboxyl groups in the interior of the pores (Sára *et al.* 1991). Since in native S-layer lattices the pores were not capable of binding strongly charged macromolecules small enough to pass through, it was concluded that the amino and carboxyl groups neutralise themselves by forming direct electrostatic interactions as was observed for the outer face of the S-layer lattice. This data clearly showed that the S-layers from the organisms investigated so far function as relatively neutral hydrophilic molecular-sieves. In the case of the S-layer glycoproteins the carbohydrate residue was also found to be exposed to the ambient environment (Messner *et al.* 1990; Sára *et al.* 1989).

S-layers as Ultrafiltration Membranes

For the production of S-layer ultrafiltration membranes (SUM) cell wall fragments or isolated S-layers were deposited on microfiltration membranes using a pressure dependent procedure and the S-layer protein was crosslinked with glutaraldehyde (Sára and Sleytr 1987c). In these composite membranes the active filtration layer consists of a coherent layer of superimposed S-layer material whereas the microfiltration membrane provides the mechanical support. Up to now, SUMs have been produced from S-layers of all crystallographic types from thermophilic and mesophilic Baillaceae. Generally, SUMs made of S-layer material from different *Bacillus stearothermophilus* strains allowed free passage for myoglobin (Mr 17 000) and carbonic anhydrase (Mr 30 000), but rejected ovalbumin (M_r 43 000) to > 90% and bovine serum albumin (M_r 67 000) to >98% (see also Table 6.1).

SUMs produced from S-layer material of *B. coagulans* or from mesophilic Bacillaceae gave similar steep rejection curves but they were shifted into the lower molecular weight range. For example, SUMs made from S-layers of *B. coagulans* rejected carbonic anhydrase to at least 90%, whereas ovalbumin and bovine serum albumin were retained to almost 100%. The steep increase in the rejection curves of SUMs clearly reflected the isoporous structure of the S-layer fragments forming the active filtration layer.

Chemical Modification of S-layer Ultrafiltration Membranes

Crosslinking the S-layer protein with glutaraldehyde during the production of the SUMs leads to net negatively charged membranes due to the reaction of a considerable proportion of free amino groups. Since under physiological conditions most proteins in solution are negatively charged it is advantageous for many ultrafiltration processes to use membranes which have a net negative charge, too. Repulsive forces between the proteins in solution and the membrane surface prevent unspecific adsorption and pore blocking which causes a decay in membrane performance such as flux losses and decrease in selectivity. SUMs with a net negative surface charge showed no or negligible flux losses after filtration of solutions of ferritin, bovine serum albumin or ovalbumin which had a net negative surface charge under the applied experimental conditions (Table 6.1; see also Küpcü et al. 1991). On the other hand filtration of solutions of positively charged proteins frequently caused high flux losess. For example, the positively charged myoglobin (M_r 17 000; pI 6.8) which was small enough to pass through the pores led to flux losses of up to 60% (Table 6.1). Although such high flux losses clearly indicated adsorption of the protein molecules in the interior of the pores, the S-layer lattices showed no rejection for myoglobin. Thus it appears that the interactions between the adsorbed protein molecules and the S-layer protein domains exposed in the pore areas are reversible involving a continuous exchange between the molecules entering the pores and those already adsorbed. On the other hand flux losses were significantly lower after filtration of positively charged proteins too large to penetrate the pore areas and pass through (Sára et al. 1991). For example, filtration of solutions of the strongly positively charged polycationised ferritin (PCF) (M_r 440 000; pI 11) caused flux losses of only 10% whereas flux losses after filtration of solutions of the small cytochrome c (M_r 12 000; pI 10.8) were around 70%. Although electron microscopical methods confirmed that PCF formed a dense monolayer on the glutaraldehyde-treated S-layer surface (see also Pum et al. chapter 10 this volume), it appears that the space left between the adjacent spherical PCF molecules was large enough to allow an almost unhindered passage for the smaller water molecules.

Table 6.1. Percentile rejection (%R) and percentile flux losses of S-layer ultrafiltration membranes produced from S-layer material of thermophilic Bacillaceae after filtration of solutions of various proteins dissolved in distilled water (pH 5.5). For chemical modification carboxyl groups from the S-layer protein were activated with 1-ethyl-3-(3-dimethylaminopropyl)carbodiimide (EDC) and subsequently allowed to react with the amino groups from different nucleophiles. Such modified membranes revealed a charge neutral surface

Nucleophile	Myoglobin M_r 17 000 pI 6.8		Carbonic anhydrase M_r 30 000 pI 5.3		Ovalbumin M_r 43 000 pI 4.3		Bovine Serum Albumin M_r 67 000 pI 4.7	
	%R	% Flux loss	%R	% Flux loss	%R	% Flux loss	%R	% Flux loss
non-modified	0	60	5	70	95	4	99	7
Glycine methylester $H-CH_2-COOCH_3$	0	5	60	10	98	10	99	10
2-amino-5-methylhexane H_3C CH_3 $CH-(CH_2)_2-CH$ H_3C NH_2	30	5	85	6	98	10	100	5
2-amino-2,4-dimethylpentane H_3C CH H_3C $CH-NH_2$ H_3C CH H_3C	55	16	85	10	98	13	100	5
3-phenylpropylamine $(C_6H_5)-(CH_2)_3-NH2$	30	10	75	4	98	8	100	7

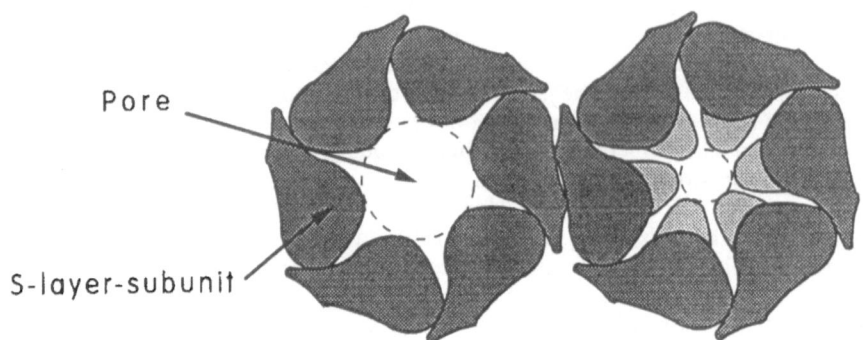

Fig. 6.1. Schematic drawing illustrating the covalent attachment of low molecular weight substances to activated carboxyl groups exposed on the S-layer protein domains in the interior of the pores. Depending on the type of substance used for immobilisation, pores of different size and shape can be formed.

Since S-layers are two-dimensional crystalline arrays composed of identical subunits, functional groups such as amino and carboxyl groups must be regularly arranged and exhibit identical positions and orientations on the individual S-layer subunits. In other words, the physicochemical properties of an S-layer are determined by those of the single constituent protein or glycoprotein subunits. This repetition of well-defined surface properties in the nanometer range makes SUMs ideal model systems for studying the effects of chemical modifications on the adsorption and rejection properties of ultrafiltration membranes.

Carboxyl groups from the S-layer protein were activated with 1-ethyl-3(3-dimethylaminopropyl)carbodiimide (EDC) and allowed to react with the free amino groups from nucleophiles of different molecular size, structure, hydrophobicity and charge. After covalent attachment of the nucleophiles glycine methyl ester (GME), 2-amino-5-methylhexane (AMH), 3-amino-2,4-dimethylpentane (ADMP) and 4,4'-3-phenyl-propylamine (PPA) SUMs with a neutral surface could be generated. Such membranes generally showed low flux losses after protein filtration (Table 6.1), but this improvement was particularly remarkable for the filtration of positively charged proteins which caused serious flux losses in non-modified membranes. By applying this carboxyl-specific modification reaction it was possible to demonstrate that, in non-modified SUMs, both free carboxyl groups in the interior of the pores as well as on the S-layer surface were responsible for protein adsorption. Further, contact angle measurements showed that, depending on the attached nucleophiles, SUMs with more hydrophilic or hydrophobic surface properties could be obtained.

Covalent attachment of low molecular weight nucleophiles to the S-layer lattice not only led to alterations in the surface properties but was also responsible for a shift of the rejection curves to the lower molecular weight range (Table 6.1). This clearly showed that carboxyl groups in the pore areas had reacted with the nucleophiles leading to an aperture-like closing of the pores. In comparison to non-modified SUMs, modified membranes showed a somewhat less steep increase in their rejection curves which can be explained by an incomplete modification of the carboxyl groups. In the case of the hexagonal S-layer lattices (Fig. 6.1, 6.2), which were used for the chemical modification studies, each pore is bordered by six identical subunits (Fig. 6.1).

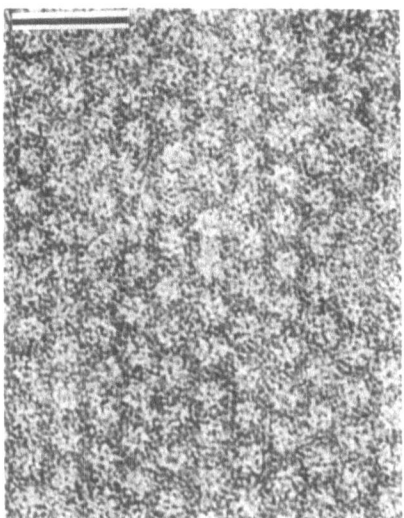

Fig. 6.2. Electron micrograph of the negatively-stained S-layer of *Clostridium thermohydrosulfuricum* L111-69. The S-layer lattice shows hexagonal symmetry. The center to center spacing of the morphological units is 14 nm. Bar, 50 nm.

If only one carboxyl group is exposed on each S-layer subunit in the interior of the pore, together six carboxyl groups would be available. Since neither the activation with EDC nor the further reaction with the nucleophiles will include all the available carboxyl groups, slightly differently sized and shaped pores can be generated. This could be responsible for the incomplete rejection of myoglobin and carbonic anhydrase showing a molecular size close to the size of the pores in the non-modified state.

To conclude, the surface and molecular-sieving properties of S-layer lattices can be specifically altered by immobilising low molecular weight compounds on activated carboxyl groups. This broad spectrum of potential modifications allows the properties of SUMs to be adapted to very specific process requirements.

Immobilisation of Macromolecules

The procedures for immobilisation of macromolecules can be classified into crosslinking, entrapping and carrier-binding which is further subdivided into physical adsorption, ionic binding and covalent binding (see also Birnbaum chapter 2 this volume; Mosbach 1987; 1988; Rosevear *et al.* 1987; Weetall and Suzuki 1975). Among the various methods, covalent binding is most frequently applied since the forces between the macromolecules and the carriers are strong and leakage of the immobilised molecules does not occur under disrupting conditions such as high salt concentrations. The carriers used for immobilising macromolecules are usually particles made of various polymers. Despite their different chemical composition all carriers reveal an amorphous structure and thus a random distribution and orientation of the functional groups. Most carriers are heteroporous and immobilisation of macromolecules can occur on the surface as well as in the interior of the particles.

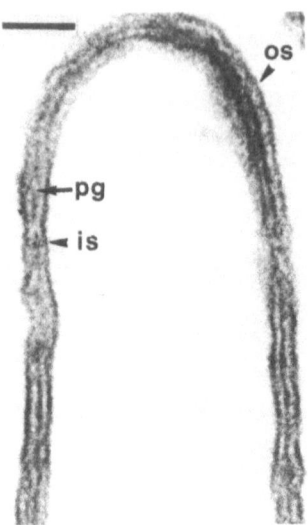

Fig. 6.3. Cup-shaped cell wall fragments as obtained from thermophilic Bacillaceae. The S-layers on both faces of the peptidoglycan-containing layer were available for the immobilisation of macromolecules. (os) outer S-layer; (pg) peptidoglycan; (is) inner S-layer. Bar, 200 nm.

With S-layer lattices immobilisation matrices with a characteristic topography and a defined arrangement and orientation of functional groups became available for the first time. These characteristic features are seen as basic requirement for an immobilisation matrix which allows a reproducible and geometrically defined binding of macromolecules.

Cup-shaped S-layer Structures as Immobilisation Matrices

To prepare cup-shaped S-layer structures as immobilisation matrices for macromolecules, whole bacterial cells from various Bacillaceae were disrupted, e.g. by ultrasonication, the cell content was removed by centrifugation and the plasma membrane was extracted with mild detergents. Cup-shaped S-layer structures are open-ended and have a size from 0.5 µm to 1 µm which is responsible for the high surface to volume ratio. They show a three-layered profile consisting of an outer S-layer, the peptidoglycan-containing layer and an inner S-layer (Fig. 6.3).

The inner S-layer is formed on the inner face of the rigid cell wall layer during the extraction of the plasma membrane from subunits not yet transferred to the cell surface (Sleytr 1978). Since the outer and the inner S-layer lattices are arranged mirror-symmetrically with respect to each other, in both cases the outer face of the S-layer lattice is accessible for immobilisation. Due to the molecular exclusion limits of S-layers from thermophilic Bacillaceae (see Table 6.1) molecules with molecular weights >40 000 can be only immobilised on the S-layer surface, but not in the interior of the pores. A further advantage of using cup-shaped S-layer structures as immobilisation matrices can be seen in the good accessibility of the immobilised macromolecules for further reactions such as enzymatic catalysis.

In order to obtain a stable immobilisation matrix the S-layer protein in the cup-shaped S-layer structures was crosslinked with glutaraldehyde. Depending on the type of organisms used for preparing cup-shaped S-layer structures, macromolecules were either directly attached to the S-layer protein or to the carbohydrate moiety in the case of S-layer glycoproteins or, if required, to spacer molecules. Spacers of different size and structure such as 4-amino butyric acid, 6-amino caproic acid, glycyl-tyrosine, poly-D-glutamic acid, poly-L- serine or poly-L-lysine which was succinylated before further activation were introduced. The carboxyl groups from the acidic amino acids of the S-layer protein or from the introduced spacer molecules were generally activated with carbodiimides (Sára and Sleytr 1989) using the procedure of Carraway and Koshland (1972). The pH applied during the activation procedure strongly depended on the type of carboxyl group and had to be optimised for each spacer. If the macromolecules were bound to the carbohydrate chains of the S-layer glycoproteins vicinal hydroxyl groups were activated with cyanogen bromide using the procedure of Axen and Ernback (1971).

In order to determine the influence of the molecular weight and the molecular size on the number of macromolecules immobilised per S-layer subunit or morphological unit, differently sized proteins were used as test molecules (Fig. 6.3). To visualise the density and binding pattern of immobilised macromolecules on the crystalline S-layer lattices, ferritin with a molecular size of 12 nm, and thus large enough to be detected by electron microscopical methods, was chosen (see also Pum *et al.* chapter 10 this volume). To determine the immobilisation capacity of S-layers as well as the retained enzymatic activity and the influence of spacer molecules on both, glucose oxidase, invertase, peroxidase and ß-D-galactosidase were used as test molecules (Sára and Sleytr 1989). The results of the immobilisation experiments on the hexagonally ordered S-layer lattice from *Clostridium thermohydrosulfuricum* L111-69 are summarised in Table 6.2.

Table 6.2. Immobilisation capacity of the S-layer lattice from *Clostridium thermohydrosulfuricum* L111-69 for ferritin (M_r 440 000) and glucose oxidase (M_r 160 000) using cup-shaped S-layer structures. The crystalline array has hexagonal symmetry. One morphological unit is composed of identical glycoprotein subunits with molecular weights of 120 000. 4-aminobutyric acid (4-ABA), 6-aminocaproic acid (6-ACA) and poly-L-lysine (PLL) which was succinylated before further activation were used as spacer molecules. Carboxyl groups from the S-layer protein or from the introduced spacer molecules were activated with 1-ethyl-3-(3-dimethylaminopropyl)carbodiimide.

	Ferritin	Glucose oxidase	
	µg/mg S-layer protein	µg/mg S-layer protein	% retained activity
-COOH	700	560	35
-COOH/4-ABA	-	520	60
-COOH/6-ACA	-	530	60
-COOH/PLL	1 000	800	50

If glucose oxidase was directly attached to the S-layer lattice 560 µg enzyme protein could be bound per mg S-layer protein. Since the molecular weight of glucose oxidase is 160 000 and that of the S-layer subunits is 120 000, the molecular ratio between glucose oxidase and S-layer subunits was 0.5. This means that 3 glucose oxidase molecules were immobilised per morphological unit of the hexagonal lattice which is composed of six identical subunits (Fig. 6.4).

Since only one of the larger ferritin molecules with a molecular weight of 440 000 and a molecular size of 12 nm was bound per morphological unit (700 µg ferritin / mg

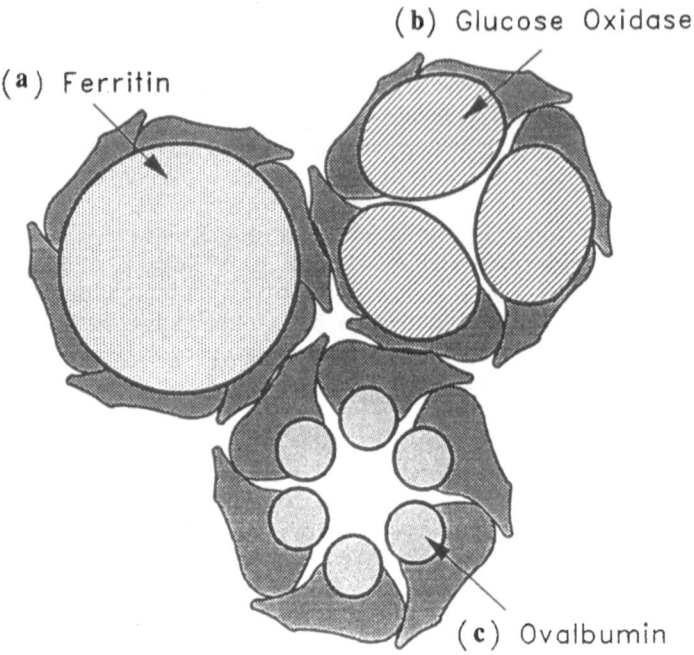

Fig. 6.4. Schematic drawing illustrating the immobilisation of differently sized macromolecules on a hexagonal S-layer lattice. Depending on the molecular weight of the macromolecules a different number could be immobilised on each morphological unit consisting of six identical subunits. (a) Immobilisation of ferritin (M_r 440 000); (b) Immobilisation of glucose oxidase (M_r160 000); (c) Immobilisation of ovalbumin (M_r 43 000).

S-layer protein), it could be demonstrated that there is a clear correlation between the number and size of macromolecules attachable per unit cell (Fig. 6.4). The immobilisation studies further showed that comparable amounts of glucose oxidase could be immobilised before and after the introduction of 4-amino butyric acid or 6-amino caproic acid as spacer molecules revealing a length of either 0.5 nm or 0.8 nm. The advantage of immobilising enzymes via spacer molecules can be seen in the higher retained enzymatic activity (Table 6.2). The highest immobilisation capacities were achieved after covalently binding poly-L-lysine (average M_r 40 000) to the S-layer lattice and succinylating the densely arranged free amino groups in the polymer chains. In the case of ferritin (molecular size 12 nm) 1.7 molecules were nominally bound per morphological unit having a size of 14 nm. Since the space occupied by a dense monolayer formed from such a high amount of ferritin is significantly larger than the space available on the S-layer surface it could be demonstrated that poly-L-lysine had enlarged the surface for immobilisation by forming an extended network on the S-layer lattice. Table 6.3 gives a survey of the immobilisation capacity of the hexagonally ordered S-layer lattice from *Clostridium thermohydrosulfuricum* after the introduction of various spacer molecules for ovalbumin (M_r 43 000).

   Ovalbumin was chosen for investigating the immobilisation density for a protein significantly smaller than the constituent S-layer subunits. If ovalbumin was directly linked to the carboxyl groups from the S-layer protein, 370 µg could be bound per mg S-layer protein which corresponds to one ovalbumin molecule per S-layer subunit (Fig. 6.4). This shows that at least one EDC-activated carboxyl group per S-layer subunit was accessible for the ovalbumin molecules with a molecular size in the range of 6 nm. Since the outer face of the S-layer lattice appeared rather corrugated in electron microscopical preparations, the presence of additional carboxyl groups not available to macromolecules due to sterical hindrance cannot be excluded. Indeed, after introduction of small spacers up to two ovalbumin molecules could be immobilised per S-layer subunit (Table 6.3).

Table 6.3. Immobilisation of ovalbumin to the S-layer lattice from *Clostridium thermohydrosulfuricum* L111-69 before and after the introduction of different spacer molecules. The S-layer lattice shows hexagonal symmetry in which one morphological unit is composed of six identical subunits with molecular weights of 120 000. Ovalbumin has a molecular weight of 43 000. Carboxyl groups from the S-layer protein or from the spacer molecules were activated with 1-ethyl-3-(3-dimethylaminopropyl)carbodiimide (EDC)

Spacer	M_r	µg Ovalbumin/mg S-layer protein	Molecular ratio Ovalbumin : S-layer subunits
no spacer	-	370	1:1
glycine	75	760	2.1:1
4-aminobutyric acid	103	620	1.7:1
6-aminocaproic acid	131	690	2:1
glutamic acid	148	590	1.6:1
poly-L-serine	5 100	370	1:1
poly-D-glutamic acid	13 600	370	1:1
poly-D-glutamic acid	41 000	490	1.4:1

To summarise, the data obtained clearly demonstrated that S-layers in comparison to the amorphous polymers usually used as carriers for covalent attachment of macromolecules represent a completely new type of immobilisation matrix. Since with cup-shaped S-layer structures the surface available for immobilisation can be derived from the amount of S-layer protein present in the suspension, it is possible to determine the ratio between immobilised macromolecules to S-layer subunits and to calculate the amount of macromolecules attached per area unit. This provides accurate information on the immobilisation density on the S-layer lattice. Furthermore, the data obtained clearly demonstrated the high binding capacity of the cup-shaped S-layer structures due to the fact that both the outer and the inner S-layer lattice were available for the immobilisaton of the macromolecules. Although in amorphous heteroporous carriers the interior of the gel matrix is also used for covalent attachment of macromolecules these matrices generally revealed lower binding capacities per gram dry weight than the crystalline S-layer lattices (Mosbach 1987).

Cup-shaped S-layer Structures as Affinity Matrices

The advantage of using S-layers as affinity matrices is seen in the possibility of attaching the ligands to a matrix with well-characterised surface properties and in the low unspecific adsorption of S-layer lattices for most proteins. To evaluate the usability of S-layer lattices as affinity matrices Protein A was linked to the S-layer protein from *Clostridium thermohydrosulfuricum* L111-69. Protein A binds specifically to the Fc region of most mammalian immunoglobulins (for review see Boyle 1990; Langone 1982a; Richman *et al.* 1982). The Fc regions of the IgG molecules are bivalent for Protein A, whereas Protein A itself has four to five potential binding sites for IgG (Moks *et al.* 1987). Thus, the complex formed between the IgGs and Protein A has a molar ratio of >2 (Deisenhofer *et al.* 1978; Hanson *et al.* 1984; Hanson and Schumaker 1984; Sjöquist *et al.* 1972). Depending on the subclasses of IgG, solubilised Protein A can bind about 8 to 12 mg human antibodies. Since the "pseudo-immune binding" is reversible in acidic environments (Duhamel *et al.* 1979; Goding 1978; MacSween and Eastwood 1981; Martin 1982), Protein A has found many applications, particularly in downstream-processing for the isolation and purification of antibodies (for review see Boyle 1990; Ey *et al.* 1978; Goding 1978; Jiskoot *et al.* 1991; Jungbauer *et al.* 1989; Langone 1982a,b; Templeton and Douglas 1978; Terman 1988; Underwood *et al.* 1983). Nowadays, two types of Protein A are commercially available. One is isolated from cell walls from *S. aureus* showing a molecular weight of approximately 42 000, the other is a recombinant protein from *E. coli* with molecular weights ranging from 32 000 (truncated Protein A) to 42 000. On the hexagonally ordered S-layer lattice of *C. thermohydrosulfuricum* L111-69 using both types of Protein A two molecules could be immobilised per S-layer subunit (M_r 120 000) (see Table 6.4). Thus, on average 12 Protein A molecules were bound per morphological unit consisting of six identical S-layer subunits (see Fig. 6.5a).

Table 6.4. Comparison of immobilising different Protein A's to the S-layer lattice from *Clostridium thermohydrosulfuricum* L111-69. The S-layer lattice shows hexagonal symmetry. One morphological unit has a diameter of 14 nm and is composed of six identical subunits with molecular weights of 120 000. Carboxyl groups from the S-layer protein were activated with 1-ethyl-3-(3-dimethylaminopropyl)carbodiimide (EDC).

	µg Protein A/mg S-layer protein	Molecular ratio Protein A : S-layer subunits	Protein A molecules per morphological unit
recombinant Protein A M_r 31 000	500	2:1	12
Staphylococcal Protein A M_r 42 000	700	2:1	12

To obtain chemically and mechanically resistant cup-shaped S-layer structures with a long shelf life, S-layers crosslinked with glutaraldehyde were subsequently treated with sodium borohydride to reduce Schiff's bases which are particularly susceptible to hydrolysis during longer periods of storage. Since S-layers with immobilised Protein A should be suitable for reusable affinity structures in downstream-processing involving crossflow conditions, Protein A was immobilised on borohydride-reduced S-layer material before and after the introduction of spacer molecules. As shown in Table 6.5 the reduced S-layer structures generally showed a lower immobilisation capacity for Protein A than the glutaraldehyde-treated S-layer lattices.

This can be explained by a loss of EDC-activated carboxyl groups due to their reaction with the secondary amino groups generated by reducing the Schiff's bases with sodium borohydride. If Protein A was directly linked to the glutaraldehyde-treated non reduced S-layer lattices, 700 µg could be bound per mg S-layer protein. This means that on average 12 Protein A molecules were immobilised per morphological unit with a size of 14 nm (Fig. 6.5a). Although on crosslinked and reduced S-layer lattices only 7 Protein A molecules could be bound to each morphological unit, both types of

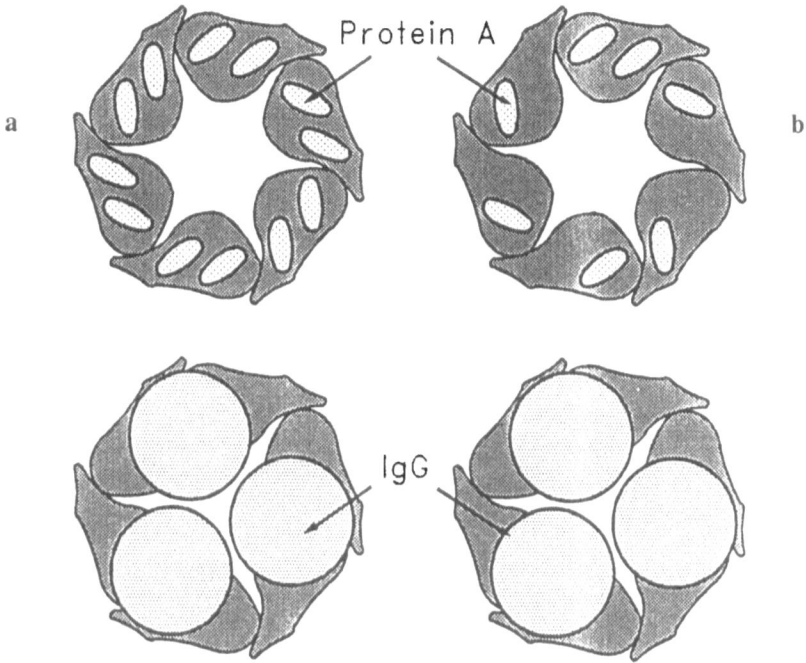

Fig. 6.5. Schematic drawing illustrating the immobilisation of Protein A (Mr 42 000) on a hexagonal S-layer lattice. (a) On S-layers crosslinked with glutaraldehyde 12 Protein A molecules could be immobilised per morphological unit, whereas (b) 7 Protein A molecules could be bound per morphological unit if the S-layer lattice was crosslinked with glutaraldehyde and reduced with borohydride. In both cases the immobilised Protein A could bind 3 IgG molecules per morphological unit of the S-layer lattice.

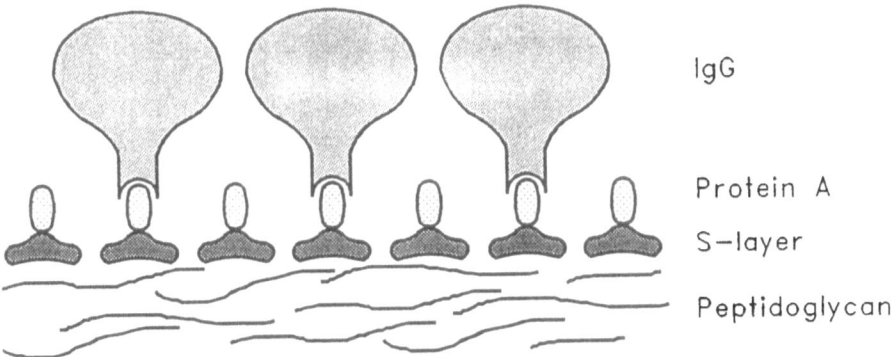

Fig. 6.6. Schematic drawing showing the immobilisation of Protein A and IgGs on the surface of an S-layer lattice. Although at least one Protein A molecule was linked to each S-layer subunit only 3 IgGs were bound per morphological unit consisting of six identical S-layer subunits (see also Fig. 6.5). This geometric consideration shows that the amount of IgG attachable per morphological unit is limited by the space available on the S-layer surface and not by the retained binding capacity of the immobilised Protein A.

Protein A carrying S-layer lattices were capable of binding 3 IgG molecules per hexameric unit cell (Fig.6.5b). In comparison to soluble Protein A, Protein A immobilised on either glutaraldehyde-treated S-layer lattices or on glutaraldehyde-treated and borohydride-reduced S-layer lattices had retained 12% and 22% of the initial IgG binding capacity, respectively (Table 6.5). The comparative Protein A immobilisation studies clearly demonstrated that it was not the amount of immobilised Protein A that was the limiting factor for IgG binding (Fig. 6.5,6.6) but the space available on the S-layer lattice. The introduction of spacer molecules had no significant influence on the amount of immobilised Protein A but led to a further reduction of the already low unspecific adsorption of the S-layer lattices and to an increase of the IgG recovery after acidic elution (Weiner unpublished observations).

Table 6.5. Immobilisation of Protein A from *S. aureus* to the S-layer lattice from *Clostridium thermohydrosulfuricum* L111-69 with and without spacer molecules using cup-shaped S-layer structures. The S-layer lattice shows hexagonal symmetry. One morphological unit showing a size of 14 nm is composed of six identical subunits with molecular weights of 120 000. glut: glutaraldehyde-treated cup-shaped S-layer structures; reduced: glutaraldehyde-treated and borohydride-reduced cup-shaped S-layer structures. Abbreviations for spacer molecules used: 4-ABA: 4-aminobutyric acid; 6-ACA: 6-aminocaproic acid; Gly: glycine Carboxyl groups from the S-layer protein were activated with 1-ethyl-3-(3-dimethylaminopropyl)-carbodiimide (EDC). The IgGs used were human polyclonal antibodies

Spacer molecules	μg Protein A/mg S-layer protein		Molecular ratio Protein A : S-layer subunits		μg IgG/mg S-layer protein		Molecular ratio IgG : Protein A		% retained activity for IgG binding	
	glut	reduced	glut	reduced	glut	reduced	glut	reduced	glut	reduced
no spacer	700	400	2:1	1,1:1	660	700	0.23:1	0.48:1	12	22
-COOH/4-ABA	630	420	5,4:1	1,2:1	670	710	0.29:1	0.46:1	13	21
-COOH/6-ACA	670	450	5,8:1	1,3:1	600	680	0.24:1	0.41:1	11	19
-COOH/Gly	-	420	-	1,2:1	-	720	-	0.47:1	-	21

Acknowledgement

This work was supported by the Fonds zur Förderung der Wissenschaftlichen Forschung in Österreich, Projekt S50/02, and the Bundesministerium für Wissenschaft und Forschung.

References

Aebi U, Smith PR, Dubochet J, Henry C, Kellenberger E (1973) Structure of the regular T-layer of *Bacillus sphaericus*. J Supramol Struct 1:498-522

Axen R, Emback S (1971) Chemical fixation of enzymes to cyanogen halide activated polysaccharide carriers. Eur J Biochem 18:351-360

Baumeister W, Engelhardt H (1987) Three-dimensional structure of bacterial surface layers. In: Harris JR, Horne RW (ed) Electron microscopy of proteins, vol 6. Academic Press, London, pp 109-154

Beveridge TJ (1981) Ultrastructure, chemistry, and function of the bacterial wall. Int Rev Cytol 72:229-317

Blatt WF (1976) Principles and practice of ultrafiltration. In: Meares P (ed) Membrane Separation Processes. Elsevier Amsterdam, pp 81-120

Burley SK, Murray RGE (1983) Structure of the regular surface layer of *Bacillus polymyxa*. Can J Microbiol 29:775-780

Carraway KL, Koshland Jr DE (1972) Carbodiimide modification of proteins. Methods Enzymol 25:616-623

Creighton TE (1983) Proteins. Structures and molecular properties. Freeman, New York

Danon D, Goldstein L, Marikovsky Y, Skutelsky E, (1972) Use of cationised ferritin as a label of negative charges on cell surfaces. J Ultrastruct Res 38:500-510

Deatherage JF, Taylor KA, Amos LA (1983) Three-dimensional arrangement of the cell wall of *Sulfolobus acidocaldarius*. J Mol Biol 167:823-852

Deisenhofer J, Jones TA, Huber R (1978) Crystallisation, crystal structure analysis and atomic model of the complex formed by a human Fc fragment and fragment B of Protein A from *Staphylococcus aureus*. Hoppe-Seyler's Z Physiol Chem 359:975-985

Dickson MR, Downing KH, Wu WH, Glaeser RM (1986) Three-dimensional structure of the surface layer protein of *Aquaspirillum serpens* VHA determined by electron crystallographic analysis. J Bacteriol 167:1025-1034

Dietz P, Hansma PK, Herrmann KH, Inacker O, Lehmann HD (1991) Atomic-force microscopy of synthetic ultrafiltration membranes in air and under water. Ultramicroscopy 35:155-159

Duhamel RC, Schur PH, Brendel K, Meezan E (1979) pH gradient elution of human IgG1, IgG2 and IgG4 from Protein A-Sepharose. J Immunol Methods 31:211-217

Engelhardt H, Saxton WO, Baumeister W (1986) Three-dimensional structure of the tetragonal surface layer of *Sporosarcina urea*. J Bacteriol 168:309-317

Ey PL, Prowse SJ, Jenkin CR (1978) Isolation of pure IgG1, IgG2a and IgG2b immunoglobulins from mouse serum using Protein A-Sepharose. Immunochem 15:429-436

Fane AG, Fell CJD, Suki A (1983) The effect of pH and ionic environment on the ultrafiltration of protein solutions with retentive membranes. J Membrane Sci 16:195-210

Goding JW (1978) Use of *Staphylococcal* Protein A as a immunological reagent. J Immunol Methods 20:241-253

Hanson DC, Phillips ML, Schumaker VN (1984) Electron microscopic and hydrodynamic studies of Protein A-IgG soluble complexes. J Immunol 132:1386-1396

Hanson DC, Schumaker VN (1984) A model for the formation and interconversion of Protein A-IgG soluble complex. J Immunol 132:1397-1409

Hovmöller S, Sjögren A, Wang DN (1988) The structure of crystalline bacterial surface layers. Prog Biophys Molec Biol 51:131-161

Jiskoot W, Van Hertrooij JJCC, Hoven AMV, Klein Gebbinck JWTM, Van der Velden-de-Grot T, Crommelin DJA, Beuvery EC (1991) Preparation of clinical grade monoclonal antibodies from serum-containing cell culture supernatants. J Immunol Methods 138:273-283

Jungbauer A, Tauer C, Wenisch E, Steindl F, Purtscher M, Reiter M, Unterluggauer F, Buchacher A, Uhl C, Katinger H (1989) Pilot scale production of a human monoclonal antibody against human immunodeficiency virus HIV-1. J Biochem Biophy Methods 19.223-240

Koval SF (1988) Paracrystalline protein surface arrays on bacteria. Can J Microbiol 34:407-414

Küpcü S, Sára M, Sleytr UB (1991) Chemical modification crystalline ultrafiltration membranes and immobilisation of macromolecules. J Membrane Sci 61:167-175

Langone JJ (1982a) Applications of immobilised Protein A in immunochemical techniques. J Immunol Methods 55:277-296

Langone JJ (1982b) Protein A of *Staphylococcus aureus* and related immunoglobulin receptors produced by *Streptococci* and *Pneumococci*. Adv Immunol 32:157-252

Lepault J, Martin N, Leonard K (1986) Three-dimensional structure of the T-layer of *Bacillus sphaericus*. J Bacteriol 168:303-308

Lonsdale H (1982) The growth of membrane technology. J Membrane Sci 10:81-95

MacSween JM, Eastwood SL (1981) Recovery of antigen from *Staphylococcal* Protein A-antibody adsorbents. Methods Enzymol 73:459-471

Martin LN (1982) Chromatographic fractionation of Rhesus monkey (*Macaca mulata*) IgG subclasses using DEAE cellulose and Protein A-Sepharose. J Immunol Methods 50:319-329

Matthiasson E (1983) The role of macromolecule adsorption in fouling of ultrafiltration membranes. J Membrane Sci 16:23-26

Messner P, Sleytr UB (1991) Bacterial surface layer (S-layer) glycoproteins. Glycobiology (in press)

Messner P, Sleytr UB (1992) Crystalline bacterial cell surface layers. In: Rose AH, Tempest DW (ed) Advances in microbial physiology, vol 33. Academic Press Inc, London (in press)

Messner P, Küpcü S, Sára M, Pum D, Sleytr UB (1990) In: Conradt HS (ed) Protein glycosylation: cellular, biotechnological and analytical aspects. CBF Monographs vol 15, VCH Weinheim, pp 111-116

Moks T, Abrahamsen L, Nilsson B, Hellman U, Sjöquist J, Uhlen M (1986) *Staphylococcal* Protein A consists of five IgG-binding domains. J Biochem 156:637-643

Mosbach K (ed) (1987) Immobilised enzymes and cells. Part C. Methods Enzymol 136, Academic Press

Mosbach K (ed) (1988) Immobilised enzymes and cells. Part D Methods Enzymol 137, Academic Press

Mulder M (1991) Basic principles of membrane technology. Kluwer Academic Publishers, Dordrecht Boston London

Pum D, Sára M, Sleytr UB, (1989a) Structure, surface charge and self-assembly of the S-layer lattice from *Bacillus coagulans* E38-66. J Bacteriol 171:5296-5303

Reihanian H, Robertson CR, Michaels AS (1983) Mechanisms of polarisation and fouling of ultrafiltration membranes by proteins. J Membrane Sci 16:237-258

Rosevear A, Kennedy JF, Cabral JMS (eds) (1987) Immobilised enzymes and cells. Adam Hilger

Sára M, Sleytr UB (1987a) Molecular sieving through S-layers of *Bacillus stearothermophilus* strains. J Bacteriol 169:4092-4098

Sára M, Sleytr UB (1987b) Charge distribution on the S-layer of *Bacillus stearothermophilus* NRS 1536/3c and the importance of charged groups for mophogenesis and function. J Bacteriol 169:2804-2809

Sára M, Sleytr UB (1987c) Production and characteristics of ultrafiltration membranes with uniform pores from two-dimensional arrays of proteins. J Membr Sci 33:27-49

Sára M, Sleytr UB (1989) Use of regulary structured bacterial cell envelope layers as matrix for the immobilisation of macromolecules. Appl Microbiol Biotechnol 30:184-189

Sára M, Moser-Thier K, Kainz U, Sleytr UB (1990) Characterisation of S-layers from mesophilic Bacillaceae and studies on their protective role towards muramidases. Arch Microbiol 153:209-214

Sára M, Pum D, Sleytr UB (1991) Permeability and charge dependent adsorption properties of the S-layer lattice from *Bacillus coagulans* E38-66. J Bacteriol (submitted)

Scherrer R, Gerhardt P (1971) Molecular-sieving by the *Bacillus megaterium* cell wall and protoplast. J Bacteriol 107:718-735

Sjöquist J, Meloun B, Hjelm H (1972) Protein A isolated from *Staphylococcus aureus* after digestion with lysostaphin. J Biochem 29:572-578

Sleytr UB, (1978) Regular arrays of macromolecules on bacterial cell walls: structure, chemistry, assembly and function. Int Rev Cytol 53:1-64

Sleytr UB, Messner P, (1983) Crystalline surface layers on bacteria. Annu Rev Microbiol 37:311-339

Sleytr UB, Messner P, (1988) Crystalline surface layers in procaryotes. J Bacteriol 170:2891-2897

Sleytr UB, Messner P (1989) Self-assembly of crystalline bacterial cell surface layers (S-layers). In Plattner H (ed) Electron microscopy of subcellular dynamics. CRC Press Boca Raton, Florida, US, pp 13- 31

Sleytr UB, Messner P, Pum D, Sára M (eds) (1988) Crystalline bacterial cell surface layers. Springer, Berlin, Heidelberg New York

Smit J (1986) Protein surface layers of bacteria. In: Inouye M (ed) Bacterial outer membranes as model systems. John Wiley & Sons, New York, USA, pp 343-376

Stewart M, Beveridge TJ (1980) Structure of the regular surface layer of *Sporosarcina urea*. J Bacteriol 142:302-309

Stewart M, Murray RGE (1982) Structure of the regular surface layer of *Aquaspirillum serpens* MW5. J Bacteriol 150:348-357

Stewart M, Beveridge TJ, Murray RGE (1980) Structure of the regular surface of *Spirillum putridiconchylium*. J Mol Biol 137:1-8

Strathmann H (1982) Membrane separation processes. J Membrane Sci 10:121-133

Templeton CL, Douglas RJ (1978) Ferritin-conjugated Protein A: a new immunocytochemical reagent for electron microscopy. FEBS Lett 85:95-98

Terman DS (1988) Preparation of Protein A immobilised on collodion-coated charcoual and plasma perfusion system for treatment of cancer. Methods in Enzymol 137:496-515

Underwood PA, Kelly JF, Harman DF, MacMillan HM (1983) Use of Protein A to remove immunoglobulins from serum in hybridoma culture media. J Immunol Methods 60:33-45

Weetall HH, Suzuki S (eds) (1975) Immobilised enzyme technology. Plenum Press

Chapter 7

Interaction of Proteins with Biomimetic Dye-Ligands: Development and Application in the Purification of Proteins

G. Kopperschläger and J. Kirchberger

Introduction

Affinity techniques exploit specific recognition phenomena being characteristic for biological systems and are ideally suited to the purification of high-value proteins. A fundamental advantage of such techniques is their predictive nature, since the ligand selected is designed to interact specifically and to possess sufficient affinity for the target protein.

A variety of ligands has found application including those exhibiting high specificity for the protein of interest like antigens for antibodies, substrates and inhibitors for enzymes, hormones for their corresponding receptor, but also such exhibiting group specificity for certain classes of proteins. The latter are called "general ligands" among which coenzymes and nucleotides, hydrophobic residues and dyes have come into widespread use.

Basically, affinity separation techniques include the following steps:

i   binding of a protein to the immobilised ligand;
ii   removal of the non-adsorbed protein; and
iii dissociation of the protein-ligand complex in order to elute the target protein by competitive effectors, by increasing or decreasing the ionic strength, by changing the pH of the buffer etc.

Selective adsorbents created with natural biological ligands are generally expensive to produce since the ligands themselves often require pre-purification, are chemically and biologically labile and tend to be difficult to immobilise to the matrices.

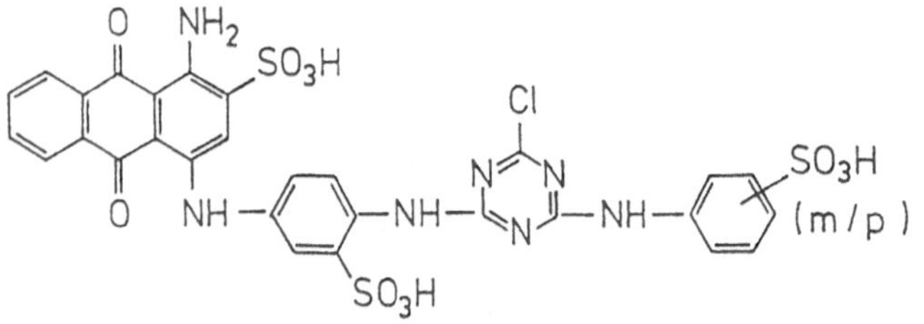

CIBACRON BLUE F3G-A

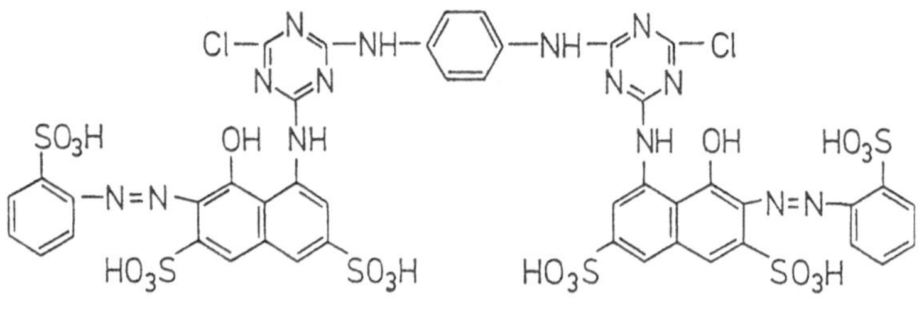

PROCION RED HE-3B

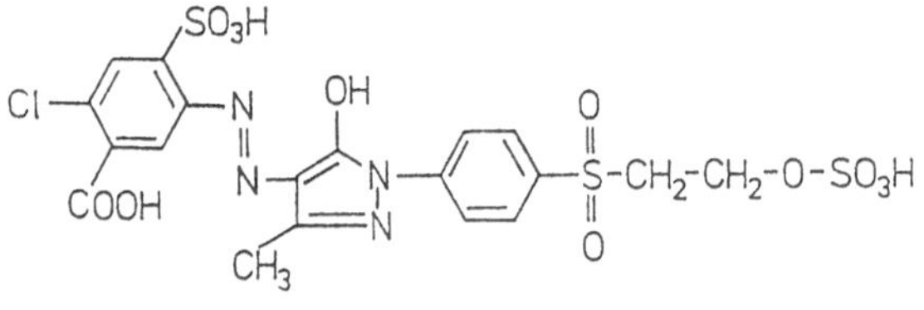

REMAZOL YELLOW GGL

Fig. 7.1. Structures of reactive dyes.

In order to circumvent these disadvantages synthetic "pseudo" or biomimetic ligands for the natural counterpart have been discovered and developed.

So far the most universal and widespread biomimetic ligands belong to the class of reactive textile dyes emanating from the original finding that Cibacron Blue F3G-A, the chromophore part of Blue Dextran, is able to bind to several kinases and dehydrogenases (Kopperschläger *et al.* 1968; Haeckel *et al.* 1968).

There are several advantages to using immobilised synthetic dyes as ligands in affinity separation techniques. Thus, they are often commercially available as low-cost chemicals in a large diversity of structures, are readily coupled to diverse matrices via their reactive group, exhibit moderate to high specificity and binding capacity, and are resistant to biological and chemical degradation.

Dye-ligands possess the potential for large-scale application in the separation technology of proteins. They have been used extensively at laboratory scale for the purification of many diagnostic, therapeutic and genetically engineered proteins (Kopperschläger et al. 1982, Clonis 1987, Scawen and Atkinson 1987).

Chemistry of Dye-Stuffs, Mode of Coupling

Basically, reactive dyes contain a specific chromophore and a reactive group which is capable of effecting a nucleophilic substitution under mild or moderate chemical conditions to several functional residues of a matrix. Both parts of the reactive dye are linked by auxochromic groups in a variety of chemical combinations.

Depending on the chemical nature of the chromophore there are anthraquinone dyes including the well known Cibacron Blue F3G-A (Ciba Geigy), azo-dyes like Procion Red HE-3B (ICI), phthalocyanine dyes like Procion Green H-4G (ICI) and dye-metal complexes like Procion Brown MX-5BR (ICI). Triazine dyes contain mono- or dichlorosubstituted triazine rings as the reactive groups whereas the class of Remazol dyes (Hoechst) reacts via the vinylsulphone group with the matrix of choice. Dichlorotriazine dyes are significantly more reactive than vinylsulphone and monochlorotriazine dyes. The chemical structures of representative dyes are shown in Figure 7.1.

Dye stuffs as marketed for commercial use normally contain various additivies to ensure adequate operational storage and handling properties. Monochlorotriazine dyes are able to withstand long periods of storage without concomitant hydrolysis, whilst dichlorotriazine dyes are more labile. Usually, the dye stuffs contain salt, traces of surface active agents and are chromophorically heterogeneous. For preparative applications highly purified dyes are not normally required since most of the contaminations do not bind covalently to the support matrices. However, for analytical applications the exclusive use of homogeneously purified dyes is recommended. For their purification preparative thin-layer chromatography, high performance liquid chromatography or column chromatography on Sephadex LH 20 should be used (Lowe and Pearson 1984).

As a rule, the dye-ligand is bound in aqueous alkaline solution directly to water-insoluble and -soluble matrices such as cross-linked agarose, cellulose, perfluorocarbons, dextran, polyethylene glycol or more rigid material like silica beads. (Lowe and Pearson 1984; Johansson and Joelsson 1985a). The introduction of a spacer between the ligand and the resin often increases both binding specificity and capacity (Lowe 1984; Naumann et al. 1989). Both, free and immobilised dyes are available individually and as kits from a variety of suppliers at modest costs.

Techniques for Studying Dye-Protein Interactions

Although an enormous number of applications of reactive dyes has been published, the chemical basis of their interaction with proteins is not yet understood in detail. Thus, in addition to the mass of empirical data derived from actual protein purifications an increasing number of direct studies of dye-protein interaction has been published in the past. Selected techniques using dyes in free as well as in immobilised form are discussed below. In addition, affinity electrophoresis (Horejsi and Ticha 1986), equilibrium dialysis (Birkenmeier and Kopperschläger 1987) and ultracentrifugation

(Cordes *et al.* 1987) have been used to obtain qualitative and quantitative data on the dye-protein interaction.

Analytical Affinity Chromatography

In addition to preparative applications, affinity chromatography can be used as an analytical tool to determine binding constants of ligand-protein interactions and to get information on the chemical nature of the binding process. The procedures are frontal and zonal elution chromatography as well as gradient elutions using competitive effectors. As shown by frontal and zonal elution chromatography of lactate dehydrogenase (LDH) and alcohol dehydrogenase (ADH) on Reactive Blue 2-Sepharose columns the dissociation constants obtained with both methods are identical and are similar to the values obtained with other methods using the free dye (Liu *et al.* 1984; Liu and Stellwagen 1986).

However, the results with respect to the binding stoichiometry were interpreted carefully because the degree of dye substitution of the matrix was found to interfere the binding behaviour.

On the other hand, the results of competitive elution experiments agreed with the finding that the dye is bound at the NADH-binding site of dehydrogenases.

Affinity chromatography has also been applied to screen a large number of reactive dyes for their biomimetic properties (Scopes 1986; Kroviakowski *et al.* 1988). Sets of columns prepacked with immobilised dyes are distributed by certain manufacturers. In addition, the HPLC-technique has been used for dye-ligand affinity chromatography by binding the dye covalently to silica beads or to other non-compressible supports (Clonis and Small 1987).

Affinity Partitioning

Dye-ligand affinity partitioning represents a special approach in the field of aqueous two-phase technology. It combines the property of biological macromolecules to partition in aqueous two-phase systems with the principle of biorecognition. As a rule, the dye-ligand is bound to one of the phase-forming polymers, usually to polyethylene glycol. Therefore, the dye-ligand partitions in a rather extreme manner in favour of one of the two phases. If a protein, being preferentially located in the opposite phase in the absence of the dye-ligand, exhibits any affinity to the dye-ligand used, the formation of the dye-protein complex leads to a change of the distribution of the protein. Under extreme conditions the target protein is completely transferred from one phase into the other. The principle of formation of aqueous two-phase systems and the effect of affinity partitioning is displayed in Figure 7.2.

The effect of affinity partitioning can be quantified by the difference of the partition coefficients of the protein obtained in the presence and in the absence of the dye-liganded polymer given on a logarithmic scale.

The parameter $\Delta \log K$ plotted against the concentration of the immobilised dye-ligand mostly follows a saturation function from which $\Delta \log K_{max}$ (maximum extraction power) and the half saturation point, $0.5 \times \Delta \log K_{max}$, can be calculated. The parameter $\Delta \log K_{max}$ is related to the number of binding sites whereas $0.5 \times \Delta \log K_{max}$ represents a relative measure of the affinity between the dye-ligand and the protein. As

two-phase system: 5% PEG^{6000}
7% Dextran T 500
50 mM sodium phosphate, pH 7.0

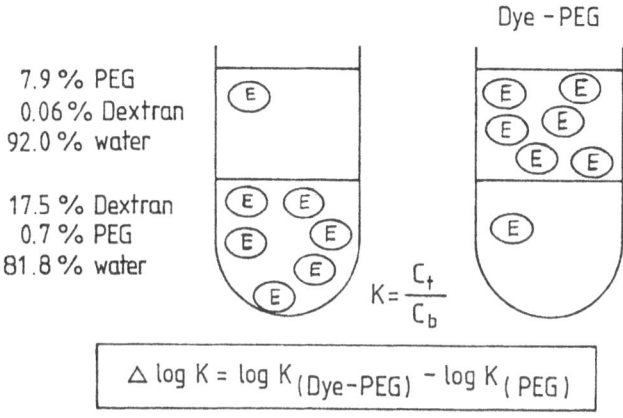

$$\Delta \log K = \log K_{(Dye-PEG)} - \log K_{(PEG)}$$

Fig. 7.2. Principle of affinity partitioning. Two-phase systems are formed by mixing, for example, 5% (w/w) polyethylene glycol 6000 (PEG 6000) and 7% (w/w) dextran T 500 in water. The upper phase contained most of the PEG 6000, the lower phase predominantly dextran as shown in Figure 7.2. Most of the proteins partition under the conditions selected in favour of the lower phase, yielding a low partition coefficient K, which expresses the ratio of the concentration of the protein in the top (C_t) and bottom phase (C_b). Upon replacement of a part of PEG 6000 by dye-liganded PEG 6000 those proteins possessing affinity to the dye-ligand are transferred from the bottom into the top phase as indicated. The parameter $\Delta \log K_{max}$ has found to be a measure for the efficiency of affinity partitioning.

an example, these two parameters are displayed in Table 7.1 for the interaction of lactate dehydrogenase from rabbit muscle with various triazine dyes (Kirchberger et al. 1989).

Table 7.1. Affinity partitioning of lactate dehydrogenase using different dye-PEG derivatives

Dye-PEG	$\Delta \log K_{max}$	Dye-PEG (μM) yielding 0.5 x $\Delta \log K_{max}$
Procion Red HE-3B	2.08	9.5
Red HE-3B(M1)	2.27	67.1
Red HE-3B(M2)	2.53	23.5
Vilmax Dye II	2.63	36.4
Cibacron Blue F3G-A	2.32	10.4
Procion Yellow HE-3G	2.63	7.2

The systems (4 g) consists of 9% (w/w) dextran 500, 6% (w/w) PEG 6000, 50 mM sodium phosphate buffer, pH 7.5 and 10 units of enzyme. The systems were equilibrated at 25°C.

There are several approaches to quantifying affinity partitioning in terms of the number of ligand binding sites and of the dissociation constant of the dye-protein complex (review see Baskir *et al.* 1989).

A general equation developed by Brooks *et al.* 1985:

$$K_c = K_p \left[1 + (C_L / k_d)_1\right]^n / \left[1 + (C_L / k_d)_2\right]^n$$

describes the dependence of the partition coefficient of the protein-dye polymer complex (K_c) on a set of parameters, where (C_L) is the concentration of the free dye polymer in the upper (1) and in the lower (2) phase, respectively, $(k_d)_1$ and $(k_d)_2$ are the dissociation constants in both phases, (K_p) is the partition coefficient of the free protein and (n) indicates the number of binding sites.

This approach assumes equal binding constants for all binding sites. Since this is obviously not correct in many cases, the calculation of the number of binding sites from affinity partitioning experiments is questionable.

Kinetic Studies

The ability of reactive dyes to act as reversible inhibitors of enzymes has been well known since 1975. Both, competitive and other forms of inhibition have been found. Especially dehydrogenases and kinases are inhibited competitively by dyes with respect to their nucleotide substrates. The strength of interaction is often significantly higher than that observed with the natural substrates. With a few exceptions, inhibition constants (K_i) are less than 10 μM (Lowe 1984).

Dyes with high reactivity, e.g. Procion MX (dichlorotriazine), generate under weak alkaline conditions a time-dependent irreversible inactivation of enzymes caused by covalent binding of the dye. This behaviour and the knowledge that some of these dyes mimic more or less biospecific effectors favours their application as site-directed labels (Small 1982).

Spectroscopic Studies

The changes in the spectral properties of a dye upon addition to a protein is caused by any interaction. The study of the spectral changes of a dye in different environments and in the presence of competitive ligands allows an estimation to be made of the main binding forces forming the dye-protein complex and reveals information on the dye-binding site of the protein. For example, the comparison of spectral changes of a dye obtained in the presence of sodium chloride or ethylene glycol with those of the protein-dye complex in aqueous solution permits a rough classification into hydrophobic and electrostatic interactions (Subramanian 1982). In a few cases, however, the difference spectrum shows a mixed-type interaction or is virtually non-existent.

The dissociation constant and the number of binding site(s) can be calculated by non-linear regression analysis if the dye-protein interaction shows both, an isosbestic point in the difference spectra and a hyperbolic titration curve (Kirchberger *et al.* 1987).

Another spectral technique - the measurement of induced circular dichroism (CD) as a function of wavelength - indicates conformational changes of the protein as a result of the formation of the dye binding and allows the characterisation of the molecular environment of the dye-binding site(s) (Edwards and Woody 1979; Towell and Woody 1980).

X-Ray Diffraction Analysis

The best technique to determine the molecular details of binding and of the conformational orientation of the dye-ligand is x-ray diffraction analysis of crystallised dye-protein complexes. But, as yet, only one x-ray diffraction data set of interaction of enzymes with triazine dyes, i.e. the interaction of horse liver alcohol dehydrogenase with Cibacron Blue F3G-A, has been published (Biellmann et al. 1979). The main reason for this situation is the protracted and difficult methods needed to get appropriate crystals of the proteins containing the dye-ligand, and so allowing high resolution.

Current Knowledge on the Chemical Basis of Protein-Dye Interaction

A careful analysis of the literature reveals that most of the reactive dyes, especially the anthraquinone dye Cibacron Blue F3G-A, show a competitive effect for the binding of natural ligands of proteins like nucleotides (GTP, ATP, IMP), cyclic nucleotides (cAMP, cGMP), pyridine nucleotide coenzymes (NAD+/NADH, NADP+/NADPH), flavins, folates, esters of coenzyme A (CoA), like acetyl-CoA and 3-hydroxy-3-methylglutaryl-CoA, monophosphate-esters (3-phosphoglycerate), hormones (thyroxine), bile pigments (bilirubin), amino and fatty acids. This finding supports the assumption that the polysulphonic aromatic chromophores mimic the naturally occuring biological heterocycles but do not sustain the hypothesis that these dyes bind only to proteins containing the so called "dinucleotide binding fold" (Thompson and Stellwagen 1976).

The high flexibility of the dye structure certainly contributes to its ability to orientate its aromatic, polar and anionic groups in such directions that a complementary binding to a wide variety of domains in proteins is achieved.

Dye-ligand affinity chromatography is not simply an anion exchange chromatography. The interaction of the polyanionic dyes with proteins is mostly electrostatic and in many cases potentiated and sometimes even dominated by hydrophobic effects.

To date, almost all studies and applications have been performed with dyes which are available from suppliers. The specificity of the dyes to act as affinity ligands, however, could be in principle improved by designing synthetic dyes which mimic more precisely the binding behaviour of natural ligands (Lowe et al. 1990).

Specifically designed dye-ligands have been successfully utilised in the affinity chromatographic purification of horse liver ADH (Burton et al. 1990) and of calf intestinal alkaline phosphatase (Lindner et al. 1989) from crude extracts with high recovery and purity.

Recent work has shown that computer-aided ligand design is now so advanced that an analogue of Cibacron Blue F3G-A, instead of NAD+, is catalytically active with horse liver ADH (Lowe et al. 1989).

Application in the Purification of Enzymes from Crude Extracts

Affinity Chromatography

Dye-ligand affinity chromatography has found significant application in the purification of hundreds of enzymes and other proteins.

Among these proteins, there is a preferential interaction of dyes with NADH-, NADPH- and other nucleotide-dependent enzymes, such as kinases, CoA-dependent enzymes, RNA and DNA nucleases, including restriction endonucleases, polymerases, phosphodiesterases, synthetases and others. Some enzymes of interest which are not retained by immobilised reactive dyes can be bound to the affinity matrix in the presence of relatively low concentrations of cations such as Zn^{++}, Co^{++}, Ni^{++}, Mg^{++} and Mn^{++}. These proteins can be selectively eluted by addition of chelating agents to the chromatographic solvent (Hughes and Sherwood 1987).

In dependence on the most dominant binding forces the bound enzymes can be eluted by increasing the ionic strength, by changing the pH of the buffer, and by adding hydrophobic agents, respectively. The use of competitive effectors like NADH, NADPH, nucleotides and others for the elution allows an increase in the degree of purification. Elution of certain enzymes by taking advantage of the formation of ternary complexes, like the coenzyme-substrate-enzyme complex, is of interest from both the theoretical and the practical point of view, because the nucleotide concentration required for dissociating the complex is much lower than in the absence of the second substrate (Gordon and Doelle 1976; Kuan et al. 1979).

Affinity Partitioning

In addition to chromatographic procedures, affinity partitioning possesses the potential to enrich enzymes from crude material by single or multistep extraction. In comparison to other affinity separation techniques, which are intrinsically batch procedures, liquid-liquid phase systems can be performed continuously and can easily be scaled up.

In order to achieve sufficient yield and a high degree of purification, a number of parameters must be optimised:

i   affinity ligands yielding sufficient specificity of interaction and high power of extraction ($\Delta \log K_{max}$);

ii   volume ratio of the phases and the partition coefficient of the bulk protein;

iii competitive effectors being capable of dissociating the enzyme-ligand complex.

The theoretical yield (Y) of a single step extraction for the top (t) and the bottom phase (b) is given by the following equations:

$$Y_t(\%) = 100 / (1 + V_b / V_t K)$$
$$Y_b(\%) = 100 / (1 + V_t K / V_b)$$

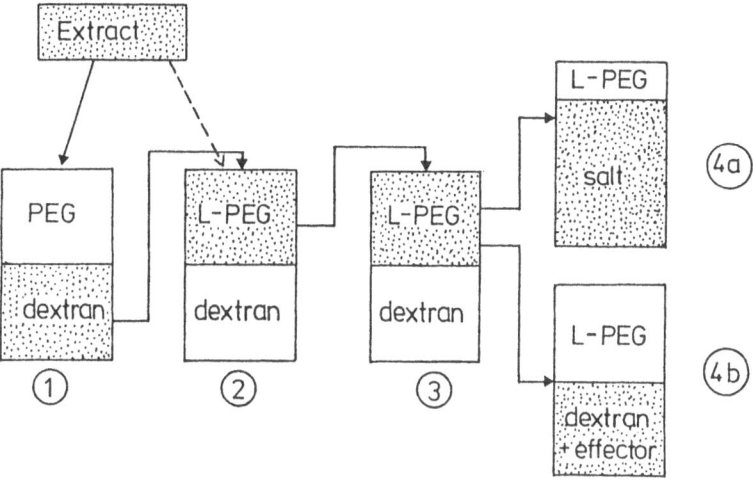

Fig. 7.3. Scheme for affinity partitioning for enzyme purification.
Step 1: Partition of an extract in a system without ligand-PEG.
Step 2: Partition of the dextran phase of step 1 or, alternatively, of the extract in the presence of ligand-PEG.
Step 3: Wash-step of the top phase of step 2 with a new bottom phase.
Step 4a: Formation of a PEG/salt two-phase system by adding solid salt to the top phase of step 3.
Step 4b: Partition of the top phase of step 3 in the presence of a new bottom phase containing a competitive effector.

One can see that both the partition coefficient (K) of a protein and the volume ratio are determining factors for the net mass transfer from one phase into the other.

Figure 7.3 shows a general scheme for employing affinity partitioning in protein purification.

It is advantageous, for removing low molecular weight material from the cell extract which often interferes with protein-ligand interaction and for concentrating the protein, to carry out precipitation of the bulk protein as a first step. Alternatively, the extract is partitioned at first in a system free of the affinity ligand in order to concentrate the bulk protein in one phase. After removal of the opposite phase a fresh one is added containing the respective dye-ligand polymer.

As a rule, affinity partitioning of crude extract cannot provide highly purified enzymes by a single step extraction. The approach has to be integrated in a series of steps involving subsequent ion exchange chromatography to remove the phase-forming polymers from the protein.

In most cases of application only one of the polymers is substituted by the ligand. However, the use of two ligands which are bound to the polymer of the top and of the bottom phase, respectively, and possessing different affinities for the target enzyme can increase the efficiency of purification (Johansson and Andersson 1984; Albertsson and Birkenmeier 1988, Pesliakas *et al.* 1988). Table 7.2 summarises enzymes which have been purified by applying affinity partitioning.

Table 7.2. Purification of enzymes by means of affinity partitioning

Enzyme	Ligand	Purification (n-fold)	Reference
Phosphofructokinase (yeast)	Cibacron Blue F3G-A	10	Kopperschläger and Johansson 1982
Glucose-6-phosphate dehydrogenase (yeast)	Procion Yellow HE-3G	22	Johansson and Joelsson 1985b
	Procion Olive MX-3G	12	
Lactate dehydrogenase (muscle)	Procion Yellow HE-3G	30	Johansson and Joelsson 1986
	Cibacron Blue F3G-A	25	
Formate dehydrogenase (*Candida boidinii*)	Procion Red HE-3B	6	Cordes and Kula 1986
3-Oxosteroid isomerase	estradiol	170	Hubert et al.1976
Alkaline phosphatase (calf intestine)	Procion Red HE-3B	13	Kirchberger (unpublished)
Alcohol dehydrogenase (yeast)	Light Resistant Yellow 2KT-Cu^{++}	6	Pesliakas et al.1988

In addition, affinity partitioning has turned out to be a simple and highly sensitive screening method for the study of dye-protein interaction (Kopperschläger et al. 1983; Birkenmeier et al. 1987; Kirchberger and Kopperschläger 1990).

Estimation of the maximum extraction power ($\Delta \log K_{max}$) allows a decision to be made as to whether an adequate interaction between an enzyme and a dye-ligand does or does not occur. However, a linear relationship to predict the efficiency of all affinity ligands has not been found. A low value of $\Delta \log K_{max}$ reflects a low efficiency of the ligand in affinity separation techniques, but a high value does not assure *per se* its usefulness in practice (Naumann et al. 1989).

Affinity Membrane Filtration

Microfiltration membranes carrying affinity ligands are potentially attractive adsorbents because they offer advantages as the matrices mentioned above together with mechanical stability, sufficient capacity, a highly porous structure and a large range of configurations for operation. A first attempt to purify enzymes from crude material was made by Champluvier and Kula (1991) by coupling a set of triazine dyes to microfiltration membranes of the Ultipor type (Pall, Dreieich, Germany). The capacity of the dye-liganded membranes was similar to the capacity of dye-liganded Sepharose and the membrane could be used repeatedly. The enrichment factor for glucose 6-phosphate dehydrogenase, malate dehydrogenase and adenylate kinase was found to be in the range of 10 starting from crude yeast extract. The membranes can be used both in batch and continuous mode.

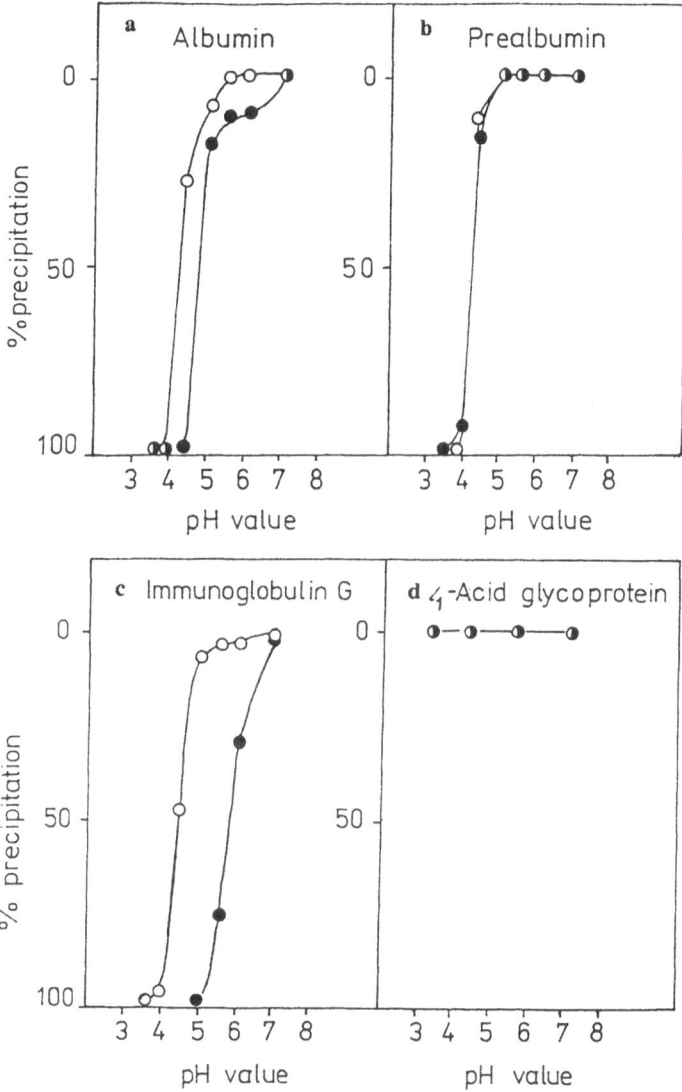

Fig. 7.4. Effect of pH on precipitation of albumin (a) prealbumin (b) immunoglobulin G (c) alpha-1-glycoprotein (d) and by Cibacron Blue F3G-A (●) and by Remazol Yellow GGL (o).

Affinity Precipitation

Based on the principle of affinity precipitation of nucleotide-dependent enzymes using bis-NAD$^+$ (Larsson and Mosbach 1979) several bifunctional bis-triazine dyes have been synthesised in order to study the precipitation behaviour of lactate dehydrogenase and albumin (Flygare *et al.* 1983; Hayet and Vijayalakshmi 1986; Pearsson *et al.* 1986). In the case of LDH optimum protein precipitation occured at a subunit:bis-dye ratio of 2:1

and was fully reversible upon addition of competitive NADH. The overall yield was more than 95% and the enrichment factor was about six yielding a highly purified enzyme (Pearson *et al.* 1986).

Besides the usefulness of bis-dyes in selective precipitation of proteins, the original dyes are also potentially capable of acting as precipitants for diverse serum proteins (Bertrand *et al.* 1985; Birkenmeier and Kopperschläger 1991).

Using albumin, prealbumin, alpha-1-acid glycoprotein and immunoglobulin G as model proteins it has been possible to demonstrate that dye-promoted precipitation depends on several factors including the structure of the dye, the pH of the solution, the dye:protein molar ratio and the intrinsic properties of the protein. In Figure 7.4 the effect of pH on affinity precipitation of albumin, prealbumin, alpha-1-glycoprotein and immunoglobulin G by two selected dyes are summarised. The efficiency of precipitation was found to increase with the complexity of the dye structure. However, the amount of dye required for complete precipitation was found to be different for any given protein (Birkenmeier and Kopperschläger 1991).

Separation of Isoenzymes

There is experimental evidence that certain biomimetic dyes are able to distinguish between multiple forms of enzymes including isoenzymes. This has been demonstrated for alkaline phosphatase from human tissues (Kirchberger *et al.* 1991a), creatine kinase from chicken heart (Wallimann *et al.* 1985), lactate dehydrogenase from rabbit tissues (Kirchberger *et al.* 1991b), collagenase from *Clostridium histolyticum* (Bond and Van Wart 1984) and aldose reductase from bovine kidney (Grimshaw 1990). For the discrimination of multiple forms of enzymes by using dye-ligands mostly kinetic and affinity chromatographic methods have been used. However, the high sensitivity in the recognition of even weak ligand-protein interactions and the simple handling also favour affinity partitioning in aqueous two-phase systems as experimental tools to study isoenzyme-dye interaction.

For example, the affinity partitioning of lactate dehydrogenase isoenzymes LDH 1 (H_4) and LDH 5 (M_4) has been studied in aqueous two-phase systems carrying the two different dye-ligands, Procion Red HE-3B and Procion Blue H-5R. The results show no difference in the affinity of LDH 5 to either dye, although the maximum affinity partitioning effect is evidently distinct, but with a significantly lower affinity of LDH 1 to Procion Blue H-5R. Affinity chromatography of both isoenzymes applying these dyes immobilised on Sepharose confirmed this behaviour (Kirchberger *et al.* 1991b).

As shown in Figure 7.5 there is no separation of LDH 1 and LDH 5 on Procion Red HE-3B-Sepharose if a gradient of NAD^+ is used. However, a complete fractionation of LDH 1 and LDH 5 can be achieved if Procion Blue H-5R is applied as an affinity ligand under the same conditions.

Plasma Protein Separation Using Dyes

The interaction of plasma proteins with immobilised Cibacron Blue F3G-A was first studied by Travis and Pannell (1973) who showed that human plasma albumin was bound to the dye, thus making albumin-depleted plasma easily available. Although

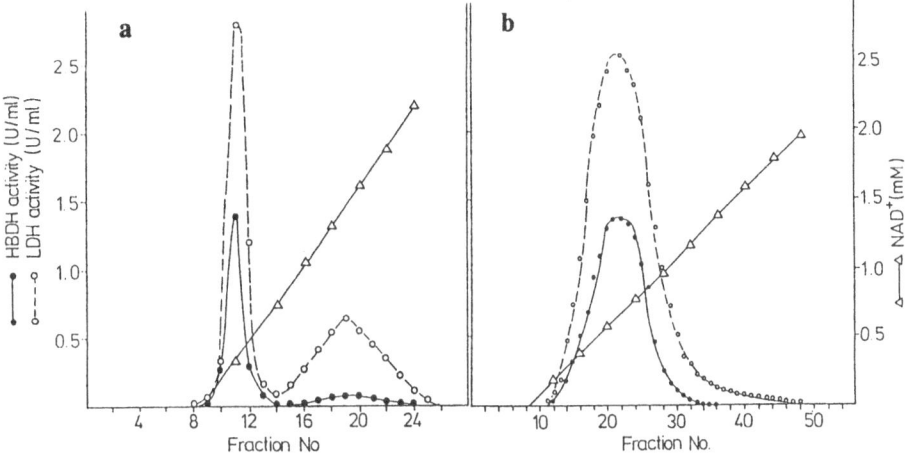

Fig. 7.5. Affinity chromatography of lactate dehydrogenase isoenzymes LDH 1 (H_4) and LDH 5 (M_4) on Procion Blue H-5R-Sepharose (**a**) and Procion Red HE-3B-Sepharose (**b**). The columns were equilibrated with 50 mM phosphate buffer, pH 7.5 at 25°C. The dialysed isoenzymes (13 units of LDH activity per each) were applied. The isoenzymes were desorbed by a linear gradient of NAD^+ (0 mM - 3 mM) in equilibration buffer. The LDH activity (-o- -o-) and the "2-hydroxybutyrate dehydrogenase activity" (-•- -•-)were determined in all fractions (1 ml).

human albumin exhibits a strong intrinsic property to interact with Cibacron Blue F3G-A other proteins also show moderate or weak interactions with the dye and can be adsorbed to affinity columns with a dependence on the degree of dye substitution (Birkenmeier and Kopperschläger 1982).

This behaviour has been successfully applied to the separation of alpha-1-fetoprotein (AFP) from human fetal material. Table 7.3 shows a purification protocol for AFP from human fetal material (Huse *et al.* 1983).

Table 3. Purification of human alpha-1-fetoprotein (AFP) from fetal material (values are the means of five typical preparations)

Procedure	Volume (ml)	Total AFP (mg)	Total protein (mg)	Yield (%)	Purification (n-fold)
Dialysis	1200	85	102 000	100	1
DEAE-cellulose chromatography	700	60	3 800	70	19
Affinity chromatography	310	35	200	41	210
Affinity rechromatography	250	30	35	35	1028
DEAE-cellulose chromatography	20	24	24	28	1200

Hemoglobin and immunoglobulins were completely removed by the first DEAE-cellulose chromatography. Using a Cibacron Blue F3G-A- Sephadex G 100 gel with a high degree of substitution (80 µg dye/mg Sephadex, dry weight) most of the proteins, with the exception of antitrypsin, transferrin and alpha-1-acid glyco–protein,

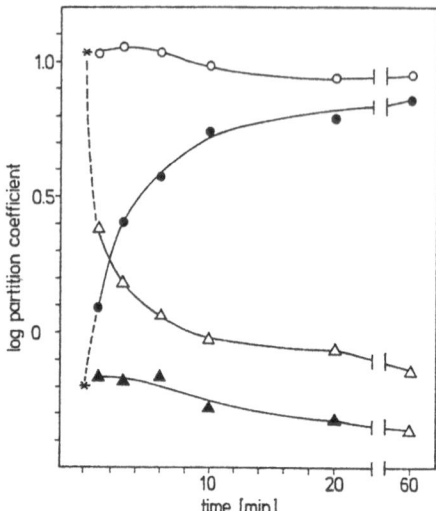

Fig. 7.6. Affinity partitioning of phosphofructokinase in the presence and in the absence of fructose 6-phosphate in dependence on the time. 0.05% of the total PEG were replaced by Procion Red HE-7B-PEG. The dialyzed enzyme was preincubated with (filled symbols) or without (open symbols) 1 mM fructose 6-phosphate. The systems contain no (circles) or 1 mM fructose 6-phosphate (triangles). (*) extrapolated values at t = 0.

were bound strongly. AFP interacts weakly with the affinity column and, hence, the retardation of this protein during washing results in separation from the break-through proteins.

Many plasma proteins are separated by taking advantage of strong affinity (albumin, lipoprotein), moderate affinity (immunoglobulins, haptoglobins, Gc-protein, antitrypsin, hemopexin) and low or non affinity (alpha-1-acid glycoprotein, alpha-2-macroglobulin, prealbumin) to immobilised dye-ligands (Birkenmeier and Kopperschläger 1989).

The use of tandem columns involving a "negative" and a "positive" chromatographic step often gives rise to an increase of the efficiency of purification. As an example, prealbumin was purified to homogeneity from human plasma by applying Cibacron Blue F3G-A-Sephadex G 100 and Remazol Yellow GGL-Sephadex G 100 as "positive" and "negative" affinity column, respectively (Birkenmeier *et al.* 1984).

Dyes as Probes for Conformational Changes of Enzymes

Reactive dyes in soluble and immobilised states have been used to recognise subtle conformational changes of proteins.

Phosphofructokinase from baker's yeast is well known for its interaction with a variety of triazine dyes (Johansson *et al.* 1983). Due to the allosteric properties of the enzyme, the substrate fructose 6- phosphate stabilises a conformation in which the inhibitory effect of ATP is relieved. Based on the fact that selected dye-ligands mimic

ATP and bind to the catalytic as well as to the regulatory sites of the enzyme, the formation and the dissociation of the enzyme-fructose 6-phosphate complex in which the regulatory site of ATP is reversibly changed can be followed by affinity partitioning as a time-dependent alteration of the partition coefficient of the enzyme which is illustrated in Figure 7.6 (Kopperschläger and Birkenmeier 1990).

The conformational change of alpha-2-macroglobulin caused by the complex formation with proteinases and accompanied by an exposure of hydrophobic sites, has been followed by affinity partitioning with Procion Yellow HE-3G-liganded PEG applying the technique of counter current distribution (Birkenmeier et. al. 1987).

As shown for *Escherichia coli* glutamine synthetase, binding of Cibacron Blue F3G-A to the relaxed, taut, oxidised and dissociated forms of the enzyme elicits different spectral perturbations of the dye and changes in the binding stoichiometry caused by different conformational states of the enzyme (Federici *et al.* 1985).

Conclusions

The versatility of reactive dyes for the purification of enzymes and other proteins is well documented. In recent years biomimetic dyes have been applied in affinity chromatography, affinity partitioning, affinity filtration and affinity precipitation. The procedures are begining to scale up to pilot plant operations in order to generate highly purified proteins for diagnostic, therapeutic and biotechnological purposes.

Dye-ligand affinity techniques for those proteins used in human therapy require careful studies on the possible biological effects of dyes (toxicity, carcinogenesis and mutagenesis), because traces of contamination in the final protein preparation caused by leakage of the affinity-ligands cannot be excluded.

However, it is believed that reactive dyes will continue to remain popular affinity-ligands in the downstream processing technology of many proteins.

References

Albertsson P-A, Birkenmeier G (1988) Affinity separation of proteins in aqueous three-phase systems. Anal Biochem 175:154-161

Baskir JN, Hatton TA, Suter UW (1989) Protein partitioning in two-phase aqueous polymer systems. Biotechnol Bioeng 34:541-558

Bertrand O, Cochet S, Kroviarski Y, Truskolaski A, Boivin P (1985) Protein precipitation induced by a textile dye. Precipitation of human plasminogen in the presence of Procion Red HE-3B. J Chromatogr 346:11-124

Biellmann JF, Samama JP, Bränden CI, Eklund H (1979) X-ray studies of the binding of Cibacron Blue F3G-A to liver alcohol dehydrogenase. Eur J Biochem 102:107-110

Birkenmeier G, Kopperschläger G (1982) Application of dye-ligand chromatography to the isolation of alpha-1-proteinase inhibitor and alpha-1-acid glycoprotein. J Chromatogr 235:237-248

Birkenmeier G, Kopperschläger G (1987) Interaction of the dye Remazol Yellow GGL to prealbumin and albumin by affinity partition, difference spectroscopy and equilibrium dialysis. Mol Cell Biochem 73:99-110

Birkenmeier G, Kopperschläger G (1989) Application of dyes in serum protein separation. Colloque INSERM 175:217-228

Birkenmeier G, Kopperschläger G (1991) Dye-promoted precipitation of serum proteins - mechanism and application. J Biotechnol 21:93-108

Birkenmeier G, Tschechonien B, Kopperschläger G (1984) Affinity chromatography and affinity partitioning of human serum prealbumin using immobilised Remazol Yellow GGL. FEBS Lett 174:162-166

Birkenmeier G, Kopperschläger G, Albertsson P-A, Johansson G, Tjerneld F, Akerlund HE, Berner S, Wickstroem H (1987) Fractionation of proteins from human serum by counter current distribution. J Biotechnol 5:115-129

Bond MD, Van Wart HE (1984) Purification and separation of individual collagenases of Clostridium histolyticum using red dye chromatography. Biochem 23:3077-3085

Brooks DE, Sharp KA, Fisher D (1985) Theoretical aspects of partitioning. In: Walter H, Brooks DE, Fisher D (eds) Partitioning in aqueous two-phase systems - theory, methods, uses, and applications to biotechnology, Academic Press, Inc., New York, pp 11-84

Burton SJ, Stead CV, Lowe CR (1990) Design and applications of biomimetic anthraquinone dyes. III. Anthraquinone-immobilised C.I. Reactive Blue 2 analogues and their interaction with horse liver alcohol dehydrogenase and other adenine nucleotide-binding proteins. J Chromatogr 508:109-125

Champluvier B, Kula MR (1991) Microfiltration membranes as pseudo-affinity adsorbents: Modification and comparison with gel beads. J Chromatogr 539:315-326

Clonis YD (1987) Large scale affinity chromatography. Bio/Technol 5:1290-1293

Clonis YD, Small DAP (1987) High-performance dye-ligand chromatography. In: Clonis YD, Atkinson T, Bruton CJ, Lowe CR (eds) Reactive dyes in protein and enzyme technology, M Stockton Press, New York, pp 87-100

Cordes A, Kula MR (1986) Process design for large-scale purification of formate dehydrogenase from Candida boidinii by affinity partitioning. J Chromatogr 376:375-384

Cordes A, Flossdorf J, Kula MR (1987) Affinity partitioning: Development of mathematical model describing behaviour of biomolecules in aqueous two-phase systems. Biotechnol Bioeng 30:514-520

Edwards RA, Woody RW (1979) Spectroscopic studies of Cibacron Blue and Congo Red bound to dehydrogenases and kinases. Evaluation of dyes as probes of the dinucleotide fold. Biochem 18:5197-5204

Federici MM, Chock PB, Stadtman ER (1985) Interaction of Cibacron Blue F3G-A with glutamine synthetase: Use of the dye as a conformational probe. 1. Studies using unfractionated dye samples.Biochem 24:647-660

Flygare S, Griffin T, Larsson P-O, Mosbach K (1983) Affinity precipitation of dehydrogenases. Anal Biochem 133:409-416

Gordon GL, Doelle HW (1976) Purification, properties and immunological relationship of L(+)-lactate dehydrogenase from *Lactobacillus caseii*. Eur J Biochem 67:543-555

Grimshaw CE (1990) Chromatographic separation of activated and unactivated forms of aldose reductase. Arch Biochem Biophys 278:273-276

Haeckel R, Hess B, Lauterborn W, Wuster K-H (1968) Purification and allosteric properties of yeast pyruvate kinase. Hoppe- Seyler's Z Phys Chem 349:699-714

Hayet M, Vijayalakshmi MA (1986) Affinity precipitation of proteins using bis-dyes. J Chromatogr 376:157-161

Horejsi V, Ticha M (1986) Qualitative and quantitative applications of affinity electrophoresis for the study of protein-ligand interaction. J Chromatogr 376:49-67

Hubert P, Dellacherie E, Neel J, Baulieu EE (1976) Affinity partitioning of steroid-binding proteins. The use of polyethyleneoxide-bound estradiol for purifying delta 5,4 3-oxoisomerase. FEBS Lett 65:169-174

Hughes P, Sherwood RF (1987) Metal ion-promoted dye-ligand chromatography. In: Clonis YD, Atkinson T, Bruton CJ, Lowe CR (eds) Reactive dyes in protein and enzyme technology, M Stockton Press, New York, pp 125-160

Huse K, Himmel M, Birkenmeier G, Bohla M, Kopperschläger G (1983) A novel purification procedure for human alpha-fetoprotein. Clin Chim Acta 133:335-340

Johansson G, Andersson M (1984) Parameters determining affinity partitioning of yeast enzymes using polymer-bound triazine dye ligands. J Chromatogr 303:39-51

Johansson G, Joelsson M (1985a) Preparation of Cibacron Blue F3G-A-poly(ethylene glycol) in large scale for use in affinity partitioning. Biotechnol Bioeng 27:621-625

Johansson G, Joelsson M (1985b) Partial purification of D-glucose 6-phosphate dehydrogenase from baker's yeast by affinity partitioning using polymer-bound triazine dyes. Enzyme Microb Technol 7:629-634

Johansson G, Joelsson M (1986) Liquid-liquid extraction of lactate dehydrogenase from muscle using polymer-bound triazine dyes. Appl Biochem Biotechnol 13:15-27

Johansson G, Kopperschläger G, Albertsson P-A (1983) Affinity partitioning of phosphofructokinase from baker's yeast using polymer-bound Cibacron Blue F3G-A. Eur J Biochem 131:589-594

Kirchberger J, Kopperschläger G (1990) An improved purification procedure of alkaline phosphatase from calf intestine by applying partitioning in aqueous two phase systems and dye-ligand chromatography. Bioseparation 1:33-41

Kirchberger J, Seidel H, Kopperschläger G (1987) Interaction of Procion Red HE-3B and other reactive dyes with alkaline phosphatase: a study by means of kinetic, difference spectroscopic and chromatographic methods. Biomed Biochim Acta 46:653-663

Kirchberger J, Cadelis F, Kopperschläger G, Vijayalakshmi MA (1989) Interaction of lactate dehydrogenase with structurally related triazine dyes using affinity partitioning and affinity chromatography. J Chromatogr 483:289-299

Kirchberger J, Domar U, Kopperschläger G, Stigbrand T (1991a) Interactions of human alkaline phosphatase isoenzymes with triazine dyes using affinity partitioning, affinity chromatography and difference spectroscopy. J Chromatogr, Biomed Appl (in press)

Kirchberger J, Kopperschläger G, Vijayalakshmi MA (1991b) Dye-ligand affinity partitioning of lactate dehydrogenase isoenzymes. J Chromatogr 557:325-334

Kopperschläger G, Johansson G (1982) Affinity partitioning with polymer-bound Cibacron Blue F3G-A for rapid, large-scale purification of phosphofructokinase from baker's yeast. Anal Biochem 124:117-124

Kopperschläger G, Birkenmeier G (1990) Affinity partitioning and extraction of proteins. Bioseparation 1:235-253

Kopperschläger G, Freyer R, Diezel W, Hofmann E (1968) Some kinetic and molecular properties of yeast phosphofructokinase. FEBS Lett 1:137-141

Kopperschläger G, Böhme H-J, Hofmann E (1982) Cibacron Blue F3G-A and related dyes as ligands in affinity chromatography. In: Fiechter A (ed) Advances in Biochemical Engineering, Springer Verlag, Berlin, Heidelberg, New York, pp 101-138

Kopperschläger G, Lorenz G, Usbeck G (1983) Application of affinity partitioning in an aqueous two-phase system to the investigation of triazine dye-enzyme interactions. J Chromatogr 259:97-105

Kroviarski Y, Cochet S, Vadon C, Truskolaski A, Boivin P, Bertrand O (1988) New strategies for the screening of a large number of immobilised dyes for the purification of enzymes. Application to the purification of enzymes from human haemolysate. J Chromatogr 449:403-412

Kuan KN, Jones GL, Vestling CS (1979) Rapid preparation of mitochondrial malate dehydrogenase from rat liver heart. Biochem 18:4366-4373

Larsson P-O, Mosbach K (1979) Affinity precipitation of enzymes. FEBS Lett 98:333-338

Lindner NM, Jeffcoat R, Lowe CR (1989) Design of applications of biomimetic anthraquinone dyes. Purification of calf intestinal alkaline phosphatase with immobilised terminal ring analogues of C.I. Reactive Blue 2. J Chromatogr 473:227-240

Liu YC, Stellwagen E (1986) Zonal chromatographic analysis of the interaction of alcohol dehydrogenase with Blue-Sepharose. J Chromatogr 376:149-155

Liu YC, Ledger R, Stellwagen E (1984) Quantitative analysis of protein-immobilised dye interaction. J Biol Chem 259:3796-3799

Lowe CR (1984) Applications of reactive dyes in biotechnology. In: Wiseman A (ed) Topics in Enzyme and fermentation biotechnology, Halsted Press, New York, pp 78-161

Lowe CR, Pearson JC (1984) Affinity chromatography on immobilised dyes. In: Jakoby WB (ed) Methods in enzyme purification and related techniques, Academic Press, London, pp 97-112

Lowe CR, Burton N, Dilmaghanian S, Mcloughlin S, Pearson J, Stewart D, Clonis YD (1989) Biomimetic dyes in biotechnology. In: Vijayalakshmi MA, Bertrand O (eds) Protein-dye interactions: Developments and applications, Elsevier Science Publ. LTD, Essex, pp 11-20

Lowe CR, Burton SJ, Burton N, Stewart DJ, Purvis DR, Pitfield I, Eapen S (1990) New developments in affinity chromatography. J Mol Recognition 3:117-122

Naumann M, Reuter R, Metz P, Kopperschläger G (1989) Affinity chromatography of bovine heart lactate dehydrogenase using dye ligands linked directly or spacer-mediated to bead cellulose. J Chromatogr 466:319-329

Pearson JC, Burton SJ, Lowe CR (1986) Affinity precipitation of lactate dehydrogenase with a triazine dye derivative: selective precipitation of rabbit muscle lactate dehydrogenase with a Procion Blue H-B analog. Anal Biochem 158:382-389

Pesliakas J-HJ, Zutautas VD, Glemza AA (1988) Affinity partitioning of yeast and horse liver alcohol dehydrogenase in polyethylene glycol-dyes/dextran two-phase systems. Chromatographia 26: 85-90

Scawen MD, Atkinson T (1987) Large-scale dye-ligand chromatography. In: Clonis YD, Atkinson T, Bruton CJ, Lowe CR (eds) Reactive dyes in protein and enzyme technology, M Stockton Press, New York, pp 51-86

Scopes RK (1986) Strategies for enzyme isolation using dye-ligand and related adsorbents.
    J Chromatogr 376:131-140
Small DAP, Lowe CR, Atkinson T, Bruton CJ (1982) Affinity labelling of enzymes with triazine dyes.
    Isolation of a peptide in the catalytic domain of horse liver alcohol dehydrogenase using Procion
    Blue MX-R as a structural probe. Eur J Biochem 128:119-123
Subramanian S (1982) Spectral changes induced in Cibacron Blue F3G-A by salts, organic solvents and
    polypeptides: Implications for blue dye interactions with proteins. Arch Biochem Biophys
    216:116-125
Thompson ST, Stellwagen E (1976) Binding of Cibacron Blue F3G-A to proteins containing the
    dinucleotide fold. Proc Nat Acad Sci USA 73:361-365
Towell JF, Woody RW (1980) Induced circular dichroism in enzyme-dye complexes: Lactic
    dehydrogenase-Bromphenol Blue. Biochem 19:4231-4237
Travis J, Pannell R (1973) Selective removal of albumin from plasma by affinity chromatography. Clin
    Chim Acta 49:49-52
Wallimann T, Zurbriggen B, Eppenberger HM (1985) Separation of mitochondrial creatine kinase
    (MiMi-CK) from cytosolic creatine kinase isoenzymes by Cibacrone-Blue affinity chromatography.
    Enzyme 33:226-231

Chapter 8

Biosensors - a Device Oriented Application of Immobilised Enzymes and Antibodies

U. Bilitewski

Introduction

Biosensors are a special group of chemical sensors consisting of a biological component and a so-called transducer (Fig. 8.1) (Hulanicki *et al.* 1989). The biological component may be any biological recognition system, but the most routinely used ones are enzymes, antibodies and microorganisms. The biological component determines the specificity of the analytical system and is usually used in an immobilised form. The support may be either the transducer, resulting in "traditional" sensors, or alternative carriers such as membranes, glass or magnetic beads, resulting in biochemical flow-through systems.

The sensitivity of the system is at least partially due to the detection principle chosen. Generally, any analytical method suitable for the determination of the reaction between the analyte and the biological component may be chosen to be the basis of the sensor. The most important methods are electrochemical, e.g. amperometric, potentiometric or conductometric measurements and optical ones, e.g. absorption and fluorescence determination. A more detailed description is given in several books and review articles dealing with principles of biosensors, (e.g. Turner *et al.* 1987a; Schmid and Karube 1988; Scheller and Schubert 1989). Characteristic features of a biosensor are its specificity, sensitivity, stability, concentration range of the analyte in which the sensor can be used, and response time. They are not only dependent on the biological component and the transducer chosen but may also be influenced by the immobilisation method. Again, various books and review articles describe immobilisation methods, (e.g. Wingard *et al.* 1976; Guilbault 1984; Hartmeier 1986), suitable for various purposes, but in the field of biosensors mainly the following principles are used:

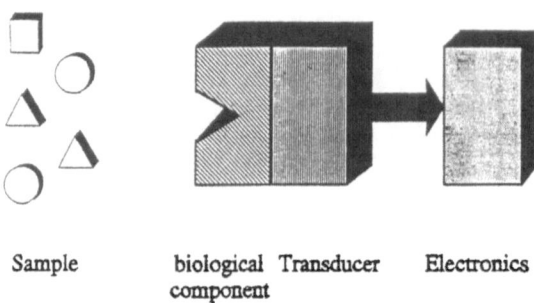

Sample   biological Transducer   Electronics
                    component

Fig. 8.1. Schematic diagram of a biosensor.

1   adsorption;
2   entrapment behind a membrane;
3   entrapment in a gel or any other suitable matrix;
4   cross-linking;
5   covalent binding to a support.

The immobilisation method has to be chosen and optimised with respect to the application. Therefore, in this article different biosensors for different applications are described with particular emphasis being placed on the importance of an optimised immobilisation method.

Biosensors Based on Enzymes

Disposable Sensors

To date oxidases are the most important group of enzymes used in disposable biosensors. The general equation of the reaction catalysed by these enzymes is

$$\text{substrate} + \text{enzyme-FAD} \rightarrow \text{product} + \text{enzyme-FADH}_2 \tag{1}$$

The enzyme is regenerated by a suitable electron acceptor which is typically molecular oxygen, O_2:

$$\text{enzyme-FADH}_2 + O_2 \rightarrow \text{enzyme-FAD} + H_2O_2 \tag{2}$$

But some enzymes are known where the oxygen may be replaced by an artificial electron acceptor, a so-called mediator (med) (Cardosi and Turner 1987):

$$\text{enzyme-FADH}_2 + \text{med(ox)} \rightarrow \text{enzyme-FAD} + \text{med(red)} \tag{3}$$

The most important mediators are ferrocene and its derivatives (Cass *et al.* 1984) and tetrathiafulvalene (Turner *et al.* 1987b).

Disposable sensors described in the literature are based on the electrochemical oxidation of either H_2O_2 or the mediator (e.g. Higgins *et al.* 1984; Foulds and Lowe 1988; Mullen 1989; Rüger *et al.* 1991; Urban *et al.* 1991; Schalkhammer *et al.* 1991;

Bilitewski *et al.* 1992a) (Equations 4, 5) or on the electrochemical reduction of oxygen (Karube and Suzuki 1988; Suzuki *et al.* 1991) (Equation 6).

$$H_2O_2 \rightarrow O_2 + 2 H^+ + 2 e^- \qquad (4)$$
electrode potential: 600 mV - 800 mV vs. Ag/AgCl with Pt-electrodes

$$med(red) \rightarrow med(ox) + e^- \qquad (5)$$
electrode potential dependent on the mediator used

$$O_2 + 2 H^+ + 2 e^- \rightarrow H_2O_2 \qquad (6)$$
electrode potential: -800 mV vs. Ag/AgCl with commercially available oxygen electrodes

The electrodes used are either conventional noble metal (Pt, Au) or carbon electrodes (Higgins *et al.* 1984; Mullen 1989) or they are fabricated by thick film (Foulds and Lowe 1988; Rüger *et al.* 1991; Bilitewski *et al.* 1992a) or thin film technology (Karube and Suzuki 1988; Urban *et al.* 1991; Schalkhammer *et al.* 1991; Suzuki *et al.* 1991). In the case of mediator-based biosensors, carbonaceous electrode materials are preferred due to the ease in modification by adsorption of compounds which are only poorly water soluble. Of the various techniques used for the immobilisation of an enzyme on the working electrode of an electrochemical set-up co-crosslinking with bovine serum albumine (BSA) and glutaraldehyde as a bifunctional reagent is probably the most common (Rüger *et al.* 1991; Suzuki *et al.* 1991; Bilitewski *et al.* 1992a). The enzyme is dissolved in a solution containing BSA, glutaraldehyde (GA) is added and this mixture is cast on the sensor. The enzyme membrane may be formed at room temperature or at 4°C with the reaction time depending on the temperature ranging from one hour to several hours (overnight). The basic reaction is

$$protein\text{-}NH_2 + CHO\text{-}(CH_2)_3\text{-}CHO + H_2N\text{-}protein \rightarrow$$
$$protein\text{-}N{=}CH\text{-}(CH_2)_3\text{-}CH{=}Nprotein \qquad (7)$$

The NH_2-groups of the protein are mainly introduced by the ε–aminogroups of lysine-residues.

Table 8.1. Optimisation of the immobilisation procedure for a glutamate-electrode (U Bilitewski 1991, unpublished data)

Composition of the enzyme/BSA/GA-solution [%]			volume applied to the electrode (µl)	linear range [µmol/l]	sensitivity [mA/mol/l]	response time (min)	cross-reactivity to glutamine
Enzyme	BSA	GA					
1.29	2.2	0.34	4*			0.5	
1.34	2.2	0.1	4*	10 - 80	0.83	0.5	
1.34	2.2	0.11	10#	0 - 70	4	3	++++
0.67	2.2	0.11	10#	20 - 70	2.4	0.5	+++
0.33	2.2	0.05	10#	40 - 130	1.87	1	++
0.16	0.3	0.01	10#	10 - 50	5	1	+ (0.25 mA/Mol/l)
0.08	0.15	0.005	10#	10 - 40	4.5	0.5	

*diameter of the electrode: 1.2 mm
#diameter of the electrode: 4 mm

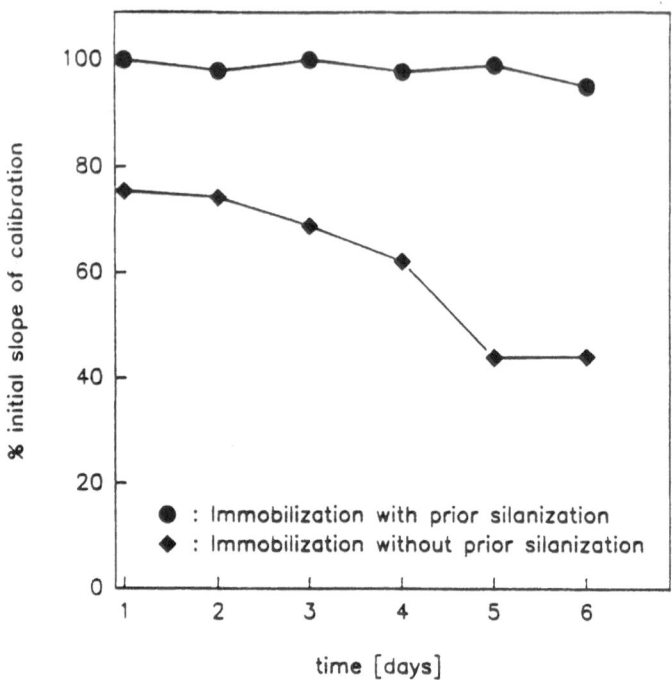

Fig. 8.2. Increased stability of an amine sensor by silanisation of the electrode.

The immobilisation procedure has to be optimised with respect to response time, sensitivity, selectivity and analytical range of the sensor, as these values depend on the thickness and porosity of the membrane and on the enzyme activity. They can be optimised by varying the concentrations of enzyme, BSA and glutaraldehyde in the immobilisation solution and by the amount of solution applied to the electrode (Table 8.1). Due to the covalent binding of the enzyme, the protein structure may be partially destroyed and hence the enzyme has a reduced specific activity compared to the native enzyme.

Some examples are described where the stability of the sensor is limited by the adhesion of the enzyme membrane to the electrode surface. To improve this the electrode may be treated with 3-amino-propyltriethoxysilane (ATS) to generate amino groups on the surface (Equation 8) to which the enzyme can be linked, again using glutaraldehyde as a coupling reagent (Equation 9):

$$\text{electrode-OH} + (C_2H_5O)_3Si(CH_2)_3NH_2 \rightarrow \text{electrode-O-Si-}(CH_2)_3NH_2 \qquad (8)$$
$$\text{electrode-O-Si-}(CH_2)_3NH_2 = \text{"electrode"-}NH_2$$

$$\text{"electrode"-}NH_2 + CHO\text{-}(CH_2)_3\text{-}CHO + NH_2\text{-protein} \rightarrow$$
$$\text{"electrode"-}N=CH\text{-}(CH_2)_3\text{-} \quad CH=N\text{-protein} \qquad (9)$$

The stability of sensors could be significantly improved by this procedure (Fig. 8.2; Bilitewski *et al.* 1992a).

Specially for the fabrication of micro-biosensors alternative immobilisation techniques are required. By electrochemical polymerisation of polypyrrole in the presence of the enzyme a polymer was formed covering exactly the surface of the working electrode. The enzyme was entrapped in this polymer (Umana and Waller 1986; Foulds and Lowe 1988; Schalkhammer *et al.* 1991) due to its pore size and by electrostatic forces. Special properties of the sensor, e.g. linear range, response time, were dependent on the conditions used for the polymerisation.

Other methods developed for the immobilisation of enzymes on thin film electrodes are the entrapment in photosensitive polymers (Shiono *et al.* 1986; Hanazato *et al.* 1987) or the covalent binding to the surface by a photosensitive compound (Urban *et al.* 1991). If these microsensors were to be applied to *in vivo* monitoring of glucose in diabetic patients the immobilisation matrix was chosen to be biocompatible (Shaw *et al.* 1991). Hence, the enzyme electrode was prepared by entrapment of the enzyme in a polyhydroxy methacrylate membrane and covering this inner membrane with an outer polyurethane membrane by a simple dip-coating procedure.

Mass production of disposable biosensors has been achieved by adapting screen printing techniques to the immobilisation of enzymes (Higgins *et al.* 1984; Mullen 1989; Bilitewski *et al.* 1992a). Pastes or inks were prepared by suspending the enzyme and platinised or palladised carbon powder in water or an organic solvent together with a suitable binder, such as hydroxyethyl cellulose, polyvinyl pyrrolidone or cellulose acetate. These pastes were applied to the working electrode by screen-printing or any other suitable, automated technology resulting in an automated production of the whole biosensor. This technique was used for the production of a glucose sensor for home care of diabetic patients, which was based on a mediated system by adding the mediator to the paste (Higgins *et al.* 1984; Matthews *et al.* 1987; Holle 1990).

As the enzyme is not covalently attached to any support and the binding is only due to adsorption at the graphite powder these biosensors do not have an extended operational stability. But they are optimised with respect to the ease of production and have an adequate shelf life when stored at 4°C.

Some of the techniques described so far for the preparation of amperometric sensors have been applied to conductivity electrodes resulting in urea sensors based on immobilised urease (Cullen *et al.* 1990; Bilitewski *et al.* 1992b).

Sensors for Continuous Analysis

Sensors for continuous determinations of analytes are integrated in automated analysers suitable for on-line monitoring of bioprocesses or for automated analysis of a high sample throughput. The basic format of these analysers is flow injection analysis (FIA). FIA was introduced in 1975 by Ruzicka and Hansen (Ruzicka and Hansen 1975) and has found widespread application to automated chemical and biochemical analysis (Schmid 1991). Basic principles are described in several books and reviews (e.g. Valcarcel and Luque de Castro (1987); Ruzicka and Hansen 1988). The sample is injected into an unsegmented continuous carrier stream, mixed with reagents and transported to a suitable detector. The specificity of such systems may be modified using immobilised enzymes which have been integrated as biosensors, mainly enzyme electrodes (Trott-Kriegeskorte *et al.* 1989), or as enzyme reactors (Schelter-Graf *et al.* 1984; Mandenius *et al.* 1985; Wehnert *et al.* 1987; Chemnitius and Schmid 1989; Kindervater *et al.* 1990; Künnecke and Schmid 1990; Spohn *et al.* 1991). Due to the application in continuous, automated analysis a high operational stability of the immobilised enzyme is required. This can be met by immobilising the enzyme in excess, such that partial denaturation or inactivation of the enzyme has only a minor influence on the signal.

Several glucose analysers and devices for other analytes based on enzyme electrodes in a flow-through cell are commercially available (Table 8.2; Scheller *et al*. 1989a). As the enzymes used are again mainly oxidases the detection is based on oxygen or hydrogen peroxide measurement. The enzyme electrodes are prepared by fabrication of an enzyme membrane and applying this to an electrode assembly. In the enzyme membrane the enzyme is entrapped in a polymer, e.g. polyurethane (Scheller *et al*. 1989b) or gelatine (Weigelt *et al*. 1987), which is placed between two membranes, e.g. dialysis membranes, cellulose acetate or polyethylene (Warsinke *et al*. 1991) membranes. The enzyme loading has been optimised by varying the amount of enzyme mixed with the polymer (Scheller *et al*. 1989b). By this procedure an operational stability of up to more than 2000 assays can be guaranteed (Table 8.2).

Table 8.2. Enzyme electrode based analysers (after Scheller *et al*. 1989a)

Company	Model	Analytes	Range of measure-ment [mmol/l]	Sample throughput [samples/h]	Stability
Yellow Springs Instr.,	23 A	Glucose	1-45	40	300 samples
Yellow Springs, OH,	23 L	Lactate	0-15	40	
USA	27	Ethanol	0-60	20	
	2700 S	Lactose			
	2700 D	Galactose	0-55	20	
		Sucrose			
		Starch	0-20 g/l	40	
		Methanol	0-60	40	
Zentrum für	Glukometer	Glucose	0.5-50	60-90	>1000 samples
Wissenschaft-		Uric Acid	0.1-1.2	40	10 d
lichen Gerätebau					
(ZWG), Berlin, FRG					
Fuji Electric, Tokio,	Gluco 20	Glucose	0-27	80-90	>500 samples
Japan		α-Amylase		30	
	UA-300A	Uric Acid		50-60	
Daiichi, Kyoto, Japan	AutoStat GA-1120	Glucose	1-40	60-120	
Radelkis, Budapest,	OP-GL-7110S	Glucose	1.7-20	40	240 d
Hungary					
Ferment, Vilnius,	ExAn	Glucose	2.5-30	20	
USSR					
La Roche, Basle,	LA 640	Lactate	0.5-12	20-30	40 d
Switzerland					
Omron Tateisi, Kyoto,	HER-100	Lactate	0-8.3		10 d
Japan					
Seres, Aix-en-	Enzymat	Glucose	0.3-22	60	
Provence, France		Choline	1.0-29	60	
		L-Lysine	0.1-2	60	
		D-Lactate	0.5-20	60	
Tacussel, Lyon,	Glucoprocesseur	Glucose	0.05-5	90	>200 samples
France					
Prüfgeräte-Werk,	ADM 300	Glucose	1-100	80	>2000 samples
Medingen, Eppendorf,	ECA 20	Glucose	0.6-60	120-130	10 d
Hamburg, FRG	ESAT 6660	Lactate	1-30	120	14 d
		Uric Acid	0.1-1.2	80	10 d

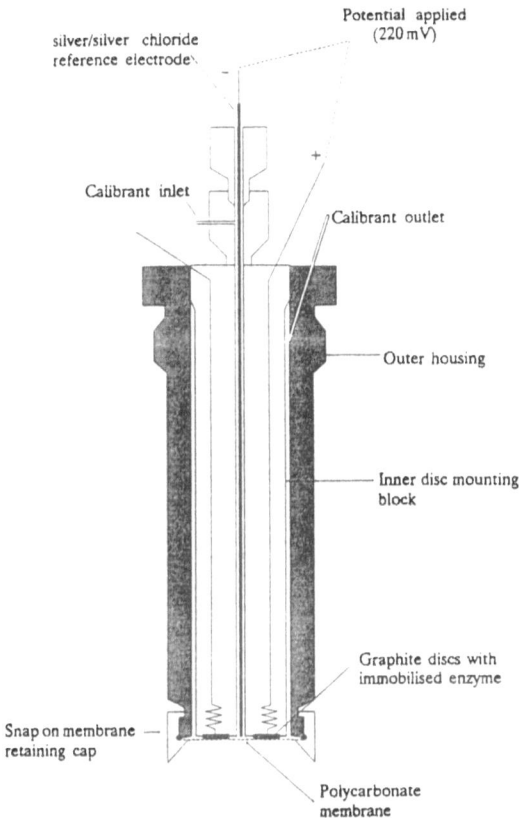

silver/silver chloride
reference electrode

Potential applied
(220 mV)

Calibrant inlet

Calibrant outlet

Outer housing

Inner disc mounting
block

Graphite discs with
immobilised enzyme

Snap on membrane
retaining cap

Polycarbonate
membrane

Fig. 8.3. Schematic vertical section of the *in situ*-glucose-fermenter electrode.

A high flexibility in the choice of a suitable enzyme and detection principle is achieved by the use of an enzyme reactor in combination with a separated detector in a flow-through system. As reagents or enzyme cofactors required for the assay may be injected together with the sample or added to the carrier, biosensor systems based on enzymes of different enzyme classes in combination with various analytical methods have been proposed in the literature. Enzyme reactors in these flow injection analysis systems consist of small polymer columns containing the support material with the immobilised enzyme. Materials typically used are nylon tubes (e.g. Roda *et al.* 1988), controlled pore glass (e.g. Masoom and Townshend 1984), oxirane-acrylic beads (Eupergit C) (e.g. Wehnert *et al.* 1985) and magnetic beads (Kindervater *et al.* 1990) with a diameter of 1-200 μm. Either they are commercially available with functional groups suitable for a covalent binding of the enzyme or they may be activated by treatment with 3-amino-propyltriethoxysilane (ATS) (see Equation 8) and glutaraldehyde (Equation 9). Due to the size of the enzyme reactors they may contain a large amount of enzyme and hence an operational stability of several weeks is reported.

Integrating methods for in-line sample pretreatment, e.g. by zone-sampling (Garn *et al.* 1988; Dremel *et al.* 1989; Chemnitius and Schmid 1989) or the use of gas-diffusion units (Künnecke and Schmid 1990; Spohn *et al.* 1991), no additional sample pretreatment was required even for on-line monitoring of bioprocesses (the sample was removed from the fermenter by sterile filtration).

Another approach to on-line monitoring of bioprocesses is the application of an in situ enzyme electrode (Bradley *et al.* 1989) comprising a sterilisable outer housing and an inner enzyme electrode (Fig. 8.3). The enzyme (glucose oxidase) was immobilised on graphite disks modified with dimethylferrocene, acting as mediator, and hexadecylamine, generating amino groups on the electrode surface. Covalent binding to the support was achieved by oxidising the mannose residues of glucose oxidase by $NaJO_4$ resulting in aldehyde groups which reacted readily with the amino groups (Brooks *et al.* 1987/88).

The stability of the electrode was limited to several days of permanent use due to leaching of the mediator. Compared to a FIA-system the time-lag between concentration changes in the fermentor and corresponding signals of the sensor was much shorter (Filippini *et al.* 1990) and data were recorded continuously, whereas with a FIA-system sampling frequencies of 15-60/h could be achieved. But the signals obtained with the in situ electrode may be influenced more strongly by changes in medium composition than signals obtained with a FIA-system, because in the FIA the sample is diluted with an optimised buffer.

Biosensors Based on Antibodies

Indirect Immunochemical Assays

Established immunochemical determinations are based on indirect assays using labelled antigens (ag) or antibodies (ab), the label being a radioisotope (RIA = radioimmunoassay), a fluorophore (FIA = fluorescence immunoassay) or an enzyme (EIA = enzyme immunoassay). Depending on the required washing steps homogeneous and heterogeneous assays are described.

The most common principle used in biosensors is the principle of an enzyme linked immunosorbent assay (ELISA). The specific antibody is immobilised on a membrane which is either fixed on an electrode (Aizawa *et al.* 1979; Robinson *et al.* 1986) or integrated in a flow injection analysis system (Krämer and Schmid 1991). The membrane may be a commercially available, already functionalised and activated one (e.g. Immunodyne, Pall Corp.) (Stöcklein *et al.* 1989) or any other support may be used which has to be activated prior to immobilisation (Aizawa *et al.* 1979) e.g. by the introduction of amino-groups or by carbodiimide (Robinson *et al.* 1986). The antibody is coupled to the membrane through its NH_2- or its COOH-groups using bifunctional reagents, e.g. butadiendiepoxide (Aizawa *et al.* 1979) or glutaraldehyde reacting in a similar way as described above for enzyme immobilisation.

By adding the sample to the antibody-membrane the antigen-antibody complex is formed. It can be determined either by a sandwich or by a competitive assay. In a sandwich assay a second, enzyme-labelled antibody is added, again forming an antigen-antibody complex with the antigen already bound to the membrane through the first antibody. The concentration of the antigen is determined by the enzyme activity bound on the membrane (Fig. 8.4). This format of an immunoassay requires large molecules with at least two antibody binding regions. For the determination of low molecular weight molecules, e.g. haptens, the format of a competitive assay is preferred using a labelled antigen competing for the binding sites of the antibody with the antigen

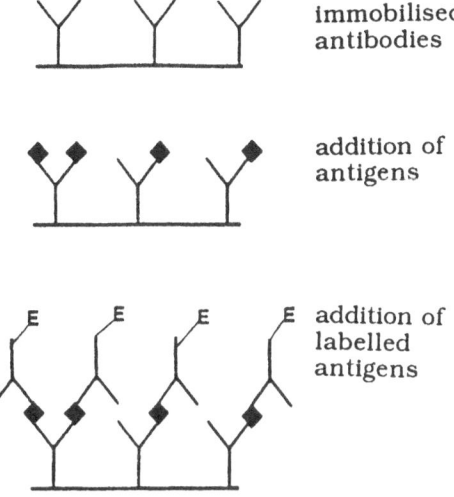

immobilised
antibodies

addition of
antigens

addition of
labelled
antigens

Fig. 8.4. Schematic diagram of a sandwich-assay.

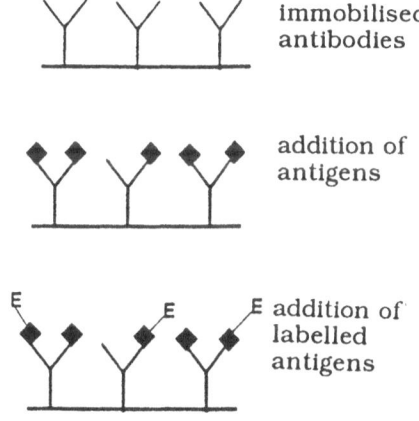

immobilised
antibodies

addition of
antigens

addition of
labelled
antigens

Fig. 8.5. Schematic diagram of a competitive assay.

Table 8.3. Comparison of a sandwich and a competitive ELISA

Sandwich ELISA	Competitive ELISA
requirement for a labelled antibody	requirement for a labelled antigen
immobilisation of the first antibody	immobilisation of the antibody
addition of the sample	addition of the sample
washing step to remove unbound antigen	washing step to remove unbound antigen
addition of the labelled antibody	addition of the labelled antigen
washing step to remove unbound antibody-enzyme-conjugate	washing step to remove unbound antigen-enzyme-conjugate
addition of the enzyme substrates	addition of the enzyme substrates
signal proportional to antigen concentration	signal inversely proportional to antigen concentration

present in the sample (Fig. 8.5). Bound enzyme activity is decreasing with increasing concentration of the antigen in the sample. Table 8.3 lists flow schemes of these two formats.

The enzyme chosen as a label should have a high specific activity to permit sensitive determinations of the antigen, e.g. horseradish peroxidase (HRP), alkaline phosphatase (AP), glucose oxidase (GOD) or catalase are used. Depending on the substrate chosen the detection principle may be either electrochemical (Aizawa *et al.* 1979; Gil *et al.* 1990; Robinson *et al.* 1986) or optical (Krämer and Schmid 1991). Suitable electrochemical determinations are the oxidation of phenol and its derivatives (AP being the label) (Gil *et al.* 1990), of mediators (GOD being the label) (Robinson *et al.* 1986) and the reduction of oxygen (catalase being the label) (Aizawa *et al.* 1979). Optical methods, i.e. measurements of absorption or fluorescence, are mainly used with HRP or AP as label.

Due to the amplification of the signal by the enzymatic reaction ELISAs may be very sensitive and allow the specific determination of analytes even in complex media as the specificity of antibodies may be rather high. But the major disadvantage is the complex procedure involving the supply of labelled antigens or antibodies and of enzyme substrates, each step separated by washing procedures to remove non-bound reactants (Table 8.3).

By establishing methods based on the evanescent field of optical fibres a simplification of immunoassays has been achieved (Dähne *et al.* 1984). The evanescent field is active at a distance outside the core of an optical fibre in the order of magnitude of the wavelength of the light. It can be used for the excitation of fluorophores which are within this distance and for collecting the emitted light into the fibre. Hence, a sandwich assay has been developed with the second antibody labelled with fluorescein isothiocyanate (FITC). The first antibody was immobilised directly on the optical fibre which was treated first with 3-amino-propyltriethoxysilane following an immersion in glutaraldehyde (Sutherland *et al.* 1984). Thus, the immobilisation procedure was similar to enzyme immobilisation on glass beads (Equations 8, 9). The antibody-containing waveguide was placed in a flow-cell to achieve an automated introduction of the sample for 10 min to form an immobilised antigen-antibody-complex. The second antibody was introduced and the reaction monitored for 10 min through the increasing fluorescence bound to the fibre through the complex "1. antibody - antigen - 2. antibody-FITC". By using this assay the number of steps could be reduced drastically.

Direct Immunochemical Assays

Following the drawbacks of indirect immunochemical assays, efforts have been made to simplify the procedure, and detection principles have been described for the direct determination of the antigen-antibody binding. Examples are the use of the piezo effect (Ngeh-Ngwainbi *et al.* 1990), of grating couplers measuring the refractive index at a glass surface (Nellen and Lukosz 1991) and of the surface plasmon resonance (Lowe *et al.* 1990). Each of these principles is sensitive to changes in the phase boundary of the transducer to its surroundings. Hence, the antibodies are immobilised directly on the transducer, typically by silanisation with ATS and covalent binding through the NH_2-groups of the antibody using glutaraldehyde as a coupling reagent (see above and Equations 8, 9), and the antigen is added. By formation of the antigen-antibody complex the mass on the surface of a piezocrystal increases which can be measured through a change in the resonance frequency of the crystal. Using grating couplers as transducers, corresponding changes in the refractive index and thickness of the surface layer can be monitored by the angle of incidence for guidance of light within the fibre or for resonance of surface plasmons within an evaporated metal layer on top of the grating.

By using these direct immunochemical formats, the assay is significantly simplified. But at the same time special care has to be taken of unspecific bindings as all effects changing the phase boundary cause a measurable signal. Another drawback may be that the amplification of the signal through an enzymatic reaction is missing and hence, the assay may be of reduced sensitivity, especially if the concentration of small molecules is to be monitored.

Conclusions

Arising from the specificity of biological recognition systems, biosensors for application in the medical and environmental field, to process monitoring and food analysis are well established in the literature. Generally, the analysis can be performed with reduced sample pretreatment, which may be integrated into an automated analyser. Though the biological component is used in an immobilised form and hence can be used for several assays, practical use of biosensors is often limited due to a limited operational stability and to the difficulty of reproducible production of biosensors. The most important step within the reproducible production of biosensors relates to the immobilisation procedure. To date the most common method is covalent attachment and cross-linking with glutaraldehyde. This may be due to the ease of this technique. Mass production is achieved only in the case of the enzyme membranes and of disposable biosensors based on screen-printing technology.

References

Aizawa M, Morioka A, Suzuki S, Nagamura Y (1979) Enzyme immunosensor. Anal Biochem 94:22-28

Bilitewski U, Chemnitius GC, Rüger P, Schmid RD (1992a) Sensors and Actuators B, accepted

Bilitewski U, Drewes W, Schmid RD (1992b) Thick film biosensors for Urea. Sensors and Actuators, accepted

Bradley J, Kidd AJ, Anderson PA, Dear AM, Ashby RE, Turner APF (1989) Rapid determination of the glucose content of molasses using a biosensor. Analyst 114:375-379

Brooks SL, Ashby RE, Turner APF, Calder MR, Clarke DJ (1987/88) Development of an on-line glucose sensor for fermentation monitoring. Biosensors 3:45-56

Cardosi MF, Turner APF (1987) In: Biosensors fundamentals and applications, eds.: Turner APF, Karube I, Wilson GS, Oxford Science Publications, Oxford University Press, 257-275

Cass AEG, Francis DG, Hill HAO, Aston WJ, Higgins IJ, Plotkin EV, Scott LDL, Turner APF (1984) Ferrocene-mediated enzyme electrode for amperometric determination of glucose. Anal Chem 56:667-71

Chemnitius GC, Schmid RD (1989) L-Malate Determinations in wines and fruit juices by flow injection analysis. Anal Lett 22 (15):2897-2913

Cullen DC, Sethi RS, Lowe CR (1990) Multi-analyte miniature conductance biosensor. Anal Chim Acta 231:33-40

Dähne C, Sutherland RM, Place JF, Ringrose AS (1984) Detection of antibody-antigen reactions at a glass-liquid interface: A novel fibre-optic sensor concept. Proc 2nd opt Fiber Conf Stuttgart 75-79

Dremel BAA, Schaffar B H, Schmid RD (1989) Determination of glucose in wine and fruit juice based on a fibre-optic glucose biosensor and flow injection analysis. Anal Chim Acta 225:293-301

Filippini C, Sonnleitner B, Fiechter A, Bradley J, Schmid R (1990) On-line determination of glucose in biotechnological processes: comparison between FIA and an in situ enzyme electrode. J of Biotechnol 18:153-160

Foulds NC, Lowe CR (1988) Immobilisation of glucose oxidase in ferrocene-modified pyrrole polymers. Anal Chem 60:2473-2478

Garn MB, Gisin M, Gross H, King P, Schmidt W, Thommen C (1988) Extensive flow-injection dilution for in-line sample pretreatment. Anal Chim Acta 207:225-231

Gil EP, Tang HT, Halsall HB, Heineman WR, Misiego AS (1990) Competitive heterogeneous enzyme immunoassay for theophylline by flow-injection analysis with electrochemical detection of p-aminophenol. Clin Chem 36/4:662-665

Guilbault GG (1984) Analytical uses of immobilised enzymes, Marcel Dekker Inc, New York and Basel

Hartmeier W (1986) Immobilisierte biokatalysatoren eine einfhrung, Springer-Verlag Berlin Heidelberg New York Tokyo

Hanazato Y, Nakako M, Maeda M, Shiono S (1987) Glucose sensor based on a field-effect transistor with a photolithographically patterned glucose oxidase membrane. Anal Chim Acta 193:87-96

Higgins IJ, McCann JM, Davis G, Hill HAO, Zwanziger R, Treidl BL, Birket NN, Plotkin EV (1984) Analytical equipment and sensor electrodes therefor. EP 0 127 958 A2

Holle W (1990) Vergleichende Beurteilung des blutglucosemeasystems exactech auf der basis von przisionsprofilen. Lab med 14:336-341

Hulanicki A, Glab S, Ingman F (1989) Chemical sensors. Definitions and classification; Vorschlge fr die IUPAC Kommission V3 (Analytische Nomenklatur)

Karube I, Suzuki H (1988) Miniaturised oxygen electrode and miniaturised biosensor and production process thereof. EP 0 284 518 A2

Kindervater R, Künnecke W, Schmid RD (1990) Exchangeable immobilised enzyme reactor for enzyme inhibition tests in flow-injection analysis using a magnetic device. Determination of pesticides in drinking water, Anal Chim Acta 234:113-117

Krämer P, Schmid RD (1991) Flow injection immunoanalysis (FIIA) - a new format for the determination of pesticides in water. Biosensors and Bioelectronics 6 (3):239-243

Künnecke W, Schmid RD (1990) Development of a gas diffusion FIA system for on-line monitoring of ethanol. J of Biotechnol 14:127-140

Lowe CR, Yon Hin BFY, Cullen DC, Evans SE, Stephens LDG, Maynard P (1990) Biosensors J Chromatogr 510:347-354

Mandenius CF, Blow L, Danielsson B, Mosbach K (1985) Monitoring and control of enzymic sucrose hydrolysis using on-line biosensors. Appl Microbiol Biotechnol 21:135-142

Masoom M, Townshend A (1984) Determination of glucose in blood by flow injection analysis and an immobilised glucose oxidase column. Anal. Chim. Acta 166:111-118

Matthews DR, Bown E, Watson A, Holman RR, Steemson J, Hughes S, Scott D (1987) Pen-sised digital 30-second blood glucose meter. The Lancet 4:778-779

Mullen WH (1989) Enzyme electrodes and improvements in the manufacture thereof. EP 0 352 925 A2

Ngeh-Ngwainbi J, Suleiman AA, Guilbault GG (1990) Piezoelectric crystal biosensors. Biosensors and Bioelectronics 5 (1):13-26

Nellen PM, Lukosz W (1991) Model experiments with integrated optical input gratimg couplers as direct immunosensors. Biosensors and Bioelectronics 6 (6):517-525

Robinson GA, Cole VM, Rattle SJ, Forrest GC (1986) Bioelectrochemical immunoassay for human chorionic gonadotrophin in serum using an electrode-immobilised capture antibody. Biosensors 2:45-57

Roda A, Girotti, Ghini S, Carrea G (1988) Methods in Enzymol. 137:161-171

Rüger P, Bilitewski U, Schmid RD (1991) Glucose and ethanol biosensors based on thick-film technology. Sensors and Actuators B 4:267-271

Ruzicka J, Hansen EH (1975) Flow injection analysis Part 1. A new concept of fast continuous flow analysis. Anal Chim Acta 78:145-157

Ruzicka J, Hansen EH (1988) Flow injection analysis. Wiley Interscience, New York

Schalkhammer T, Mann-Buxbaum E, Urban G, Pittner F (1991) Electrochemical glucosesensors on permselective nonconducting substituted pyrrolepolymers, Sensors and Actuators B, 4:

Scheller F, Schubert F (1989) Biosensoren Birkhuser Verlag Basel Boston Berlin

Scheller F, Schubert F, Pfeiffer D, Hintsche R, Dransfeld I, Renneberg R, Wollenberger U, Riedel K, Pavlova M, Khn M, Mller H-G, Pham minh Tan, Hoffmann W, Moritz W (1989a) Research and development of biosensors. A Review. Analyst 114:653-662

Scheller FW, Pfeiffer D, Hintsche R, Dransfeld I, Nentwig J (1989b) Glucose measurement in diluted blood. Biomed Biochim Acta 48:891-896

Schelter-Graf A, Schmidt H-L, Huck H (1984) Determination of the substrates of dehydrogenases in biological material in flow-injection systems with electrocatalytic NADH oxidation. Anal. Chim. Acta. 163:299-303

Schmid RD, Karube I (1988) Biosensors and "Bioelectronics", In: Biotechnology Vol 6b eds.: RehmH-J, Reed G, VCH Verlagsgesellschaft, Weinheim 317-365

Schmid RD Ed.: (1991) Flow injection analysis (FIA) based on enzymes or antibodies, VCH Weinheim New York Basel Cambridge (GBF Monographs Vol. 14)

Shaw GW, Claremont DJ, Pickup JC (1991) In vitro testing of a simply constructed, highly stable glucose sensor suitable for implantation in diabetic patients. Biosensors and Bioelectronics 6 (5):401-406

Shiono S, Hanazato Y, Nakako M (1986) Urea and glucose sensors based on ion sensitive field effect transistor with photolithographically patterned enzyme membrane. Anal. Sciences 2:517-521

Spohn U, Eberhardt R, Joksch B, Wichmann R, Wandrey C, Voß H (1991) Enzymatic multi-channel-FIA methods for on-line fermentation monitoring and control. In: Flow injection analysis (FIA) based on enzymes or antibodies, VCH Weinheim New York Basel Cambridge (GBF Monographs Vol. 14) pp 51-62

Stöcklein W, Krämer P, Schmid RD (1989) Flow injection immunoanalysis (FIIA) for the determination of pesticides in water. In: Biosensors applications in medicine, environmental protection and process control. Eds.: Schmid RD, Scheller F, VCH Weinheim (GBF Monographs Vol. 13) pp 307-312

Sutherland RM, Dähne C, Place JF, Ringrose AR (1984) Immunoassays at a quartz-liquid interface: theory, instrumentation and preliminary application to the fluorescent immunoassay of human immunoglobulin G, J Immunol Meth 74:253-265

Suzuki H, Sugama A, Kojima N, Takei F, Ikegami K (1991) A miniaturised Clark-type oxygen electrode using a polyelectrolyte and its application as a glucose sensor, Biosensors and Bioelectronics 6 (5):395-400

Trott-Kriegeskorte G, Renneberg R, Pawlowa M, Schubert F, Hammer J, Jäger V, Wagner R, Schmid RD, Scheller F (1989) Enzyme sensors for process control of cell cultures In: Biosensors application in medicine, environmental protection and process control, Eds.: Schmid RD, Scheller F, VCH Weinheim (GBF Monographs Vol. 13) pp 67-70

Turner APF, Karube I, Wilson GS Eds.: (1987a) Biosensors fundamentals and applications, Oxford Science Publications, Oxford University Press

Turner APF, Hendry SP, Cardosi MF (1987b) Application of tetrathiafulvalenes in bioelectrochemical processes EP 0 234 938 A2

Umana M, Waller J (1986) Protein-modified electrodes. The glucose oxidase/polypyrrole system. Anal Chem 58:2979-2983

Urban G, Jobst G, Kohl F, Jachimowicz A, Olcaytug F, Tilado O, Goiser P, Nauer G, Pittner F, Schalkhammer T, Mann-Buxbaum E (1991) Miniaturised thin-film biosensors using covalently immobilised glucose oxidase. Biosensors and Bioelectronocs 6(7):555-562

Valcarcel M, Luque de Castro MD (1987) Flow-injection analysis principles and applications, Ellis Horwood Limited Publishers, Chichester (Ellis Horwood Series in Analytical Chemistry)

Warsinke A, Renneberg R, Scheller F (1991) Amperometric multienzyme sensor for determination of D-glucono-k-lactone. Anal Lett 24 (8):1363-1373

Wehnert G, Sauerbrei A, Schgerl K (1985) Glucose oxidase immobilised on Eupergit C and CPG-10. A Comparison. Biotechnol Lett 7 (11):827-830

Wehnert G, Sauerbrei A, Bayer T, Scheper T, Schgerl K (1987) Application of an enzyme thermistor for the determination of glucose in complex fermentation media. Anal Chim Acta 200:73-78

Weigelt D, Schubert F, Scheller F (1987) Enzyme sensor for the determination of lactate and lactate dehydrogenase activity. Analyst 112:1155-1158

Wingard, Jr. LB, Katchalski-Katzir E, Goldstein L Eds.: (1976) Immobilised enzyme principles, Academic Press New York San Francisco London (Applied Biochemistry and Bioengineering Vol. 1)

Chapter 9

New Immobilisation Techniques for the Preparation of Thin Film Biosensors

Th. Schalkhammer, E. Mann-Buxbaum, I. Moser, G. Hawa, G. Urban and F. Pittner

Introduction

The demand for microbiosensors in medicine and biotechnology has increased dramatically in the last few years. A wide variety of enzyme-based biosensors has been developed. Electrochemical techniques offering several advantages for sensor design in that they are rapid, sensitive and easily interfaced to computers are among the main detection principles. Whereas direct electrochemical detection of most biological substances is neither selective nor able to quantify low concentrations, enzymes (mainly oxidases, dehydrogenases and ammonia liberating enzymes) are able to convert many substances into electrochemically active species.

Due to the wide field of applications, electrochemical glucose sensors are the most important and developed biosensors for the moment. Glucose oxidase, a highly stable and well characterised commercially available oxidoreductase, producing hydrogen peroxide, is the bio-component of most or nearly all glucose sensors on the market. Hydrogen peroxide oxidation as well as electron transfer employing various mediators on metal or carbon electrodes creates an amperometric signal.

However, various problems have to be solved, to allow the construction of highly satisfactory sensors for practical use in great numbers:

i   The stability of the immobilised biosystem e.g. enzyme, cofactor,.(even under dry storage conditions).
ii   Stable and efficient coupling procedures.
iii Selectivity of the electrochemical transducer steps.
iv   The mechanical stability of the sensor.
v   The reproducibility with respect to response and temperature dependence.

Thin film and thick film (Schmid *et al.* 1991) techniques offer a wide variety of possibilities in designing miniaturised electrochemical sensors. Especially thin film technology is able to provide the high purity and reproducibility required of the electrode surface and the high spatial resolution of the electrode structure.

As it is not feasible to employ macroscopic techniques for microdevices, stepwise derivatisation, starting from the metal electrode surface, has turned out to be the most practical. To reach this objective different techniques have been studied and are compared.

The outstanding difference of monolayer thin film biosensors compared to conventional types of biosensors is the crucial importance of interface and surface chemistry.

The following formula will give the amount (number of enzyme molecules) of enzyme bound in a thick film membrane biosensor:

$$Z_{Enzyme} = a^2 \, (m^2) \cdot d \, (m) \cdot c_{Enzyme} \, (mol \, l^{-1}) \cdot N/1000$$

a^2 = dimension of sensor electrode
d   = thickness of enzyme layer

The only parameter having a high degree of deviation (due to the sensor preparation) may be the thickness of the enzyme layer.

On the other hand the following formula will give the amount of enzyme coupled to a monolayer thin film biosensor:

$$Z_{Enzyme} = a^2 \, (m^2) \cdot r_c \cdot \varphi \cdot \sum_{i = 1...n} \varphi_i / ai^2 \, (m^2)$$

r_c = correction coefficient for surface roughness
φ = degree of surface coverage
φ_i = relative amount of surface occupied by protein n
a_i^2 = absolute surface area occupied by protein n

The first three parameters are mainly influenced by the type of sensor substrate and a_i^2 is constant only depending on the size of a given protein, while φ_i is the result of a very complex protein surface interaction chemistry.

Sensor

Sensor Design

Microstructuring of sensor substrates by thin film technology has already become state of art as a result of the high know-how gain in the field of microelectronics and IC design. Not only silicon and ceramic wafers but also polymeric supports (e.g. polyimide, polycarbonate, etc) can be used as sensor substrates. A typical procedure for sensor construction is given in the following paragraph :

Sodium silicate glass sheets of 0.3 mm thickness and aluminium oxide ceramics 0.6 mm were used as electrode carriers. After standard cleaning procedures with detergents and sputter etching with argon ions, metals were evaporated by an electron gun in a high vacuum instrument (Balzers) and used to coat the substrates with titanium up to a thickness of 80 nm as an adhesion layer. Platinum up to a thickness of 60 nm being evaporated on top of the titanium film acts as an electrochemical electrode. Structuring of these thin films was performed by a lift off technique with AZ

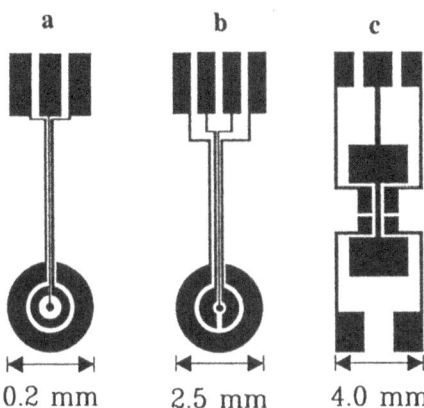

Fig. 9.1. Types of electrode used for immobilisation of proteins or DNA.

5218E photoresist. These layers were isolated by a 1000 nm siliconnitride layer and structured by plasma etching. The platinum surface was cleaned by etching with oxygen plasma (30 watt) for 3 min and if necessary the surface was hydrophobised by dipping into a 10% solution of hexamethyldisilazane in xylene for 15 min.

Using these procedures three different electrode types were constructed :

i   a 3-electrode electrochemical cell (outer diameter of 0.2 mm) possessing one working electrode. A Ag/AgCl - reference electrode was produced by evaporating and structuring a 1000 nm silver film, which was oxidised with aqueous $KMnO_4$. (1%, 2 min) and subsequently chlorinated with a solution of KCl (3 M). (see Fig. 9.1a).

ii   a 4-electrode electrochemical cell (outer diameter 2.5 mm) with two working electrodes and a Ag/ AgCl reference electrode was produced as described above (Fig. 9.1b)

iii a 5-electrode electrochemical cell (outer diameter of 4 mm) possessing three working electrodes. A Ag/AgCl - reference electrode was produced by evaporating and structuring a 1000 nm silver film, which was oxidised with aqueous $KMnO_4$.(1%, 2 min) and subsequently chlorinated with a solution of KCl (3 M) (see Fig. 9.1c.

Metal Surface

Only a small number of metals can be used to act as an electrochemical electrode for thin film biosensors. Mostly the precious metals platinum and gold but also rhodium, palladium and iridium can be used as electrode materials. For special applications thin carbonated films of polymers can be used instead of glassy carbon. Due to the high degree of surface oxidation palladium and iridium, electrochemistry is rather complex. Gold electrodes are sensitive to destruction by complexing agents. Because of that the cheaper platinum is preferred in most cases, as can be seen from the literature. Conducting glass (e.g. tin -/ indiumoxide) can also be used as an electrode for thin film

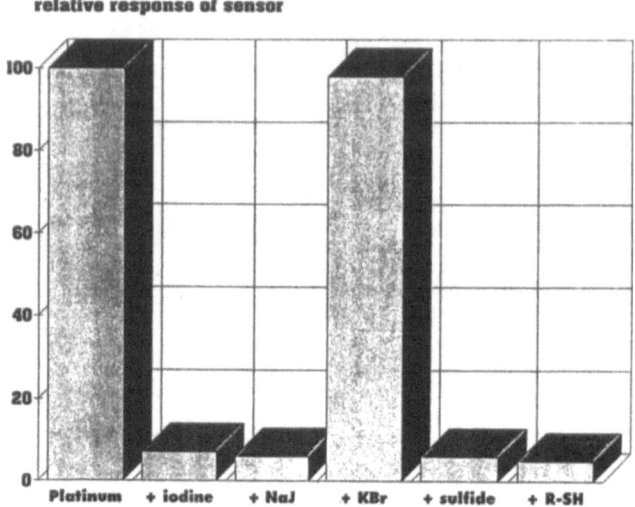

Fig. 9.2. Reversible changes in sensor response of a mercaptosilane coated sensor.

biosensors but the electron transfer at the interface and the ion mobility is lower than on metal electrodes (Okawa *et al.* 1989).

Using platinum as an electrode material, oxidation and silane treatment are most effective for metal activation. For protein immobilisation aminosilanes are most convenient, but due to the high unspecific adsorption background for DNA it should not be used for oligonucleotide immobilisation.Testing the electrochemistry of silanised sensors electrodes it turned out that the platinum surface is blocked by thiol-containing silane compounds. Thus it is necessary to modify the electrodes by oxidising the chemisorbed Pt - sulfur thereby reactivating the platinum electrode surface. Voltage pulsing and cycling up to 1300 mV/(Ag/AgCl) is necessary to remove the covalently coupled sulfur from the platinum surface .

Moreover sensors coated with a mercaptosilane layer and electrochemically cleaned as listed above showed a significant decrease in electrochemical response (> 90%) when treated with solutions of iodine, iodide, thiocyanate, sulfide and thiols in trace amounts. No effect was observed with bromide, chloride, nitrate and chlorate. Voltage pulsing and cycling in phosphate buffer pH = 7.0 up to 1200 mV (Ag/AgCl) (for iodine) and up to 1300 mV (for sulfur containing compounds) enable the platinum biosensor surface to be cleaned without destroying the layers above (Fig. 9.2).

Cationic impurities were able to modify the silane coated platinum surface. Especially AgCl which may leak from the reference electrode during prolonged exposure to aqueous solvents caused a modification of the platinum electrochemistry by forming a stable surface phase. A very intense and sharp peak could be observed as an indicator for this AgCl leakage from the reference electrode. The peak was moving and broadening when the chloride concentration was lowered.

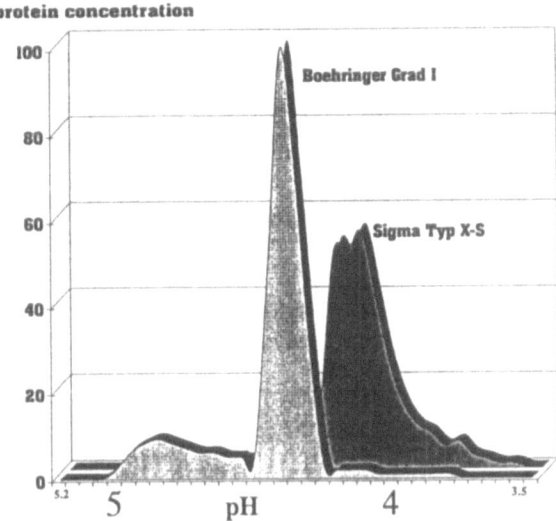

Fig. 9.3. Glucose oxidase purchased from different sources gave different characteristics when purified by chromatofocusing.

Adsorption

Adsorption of Proteins

Protein chemistry on metal surfaces is significantly more complex than adsorption and immobilisation of proteins on glass or organic polymer coated supports (Wahlgren and Arnebrant 1991, Wingard and Miyawaki 1984, Bergveld 1991). The high enthalpy of chemisorption caused by the interaction of different functional groups on the proteins with the top metal layer atoms results in a nearly irreversible coating of the metal electrodes. Chemical modification of the electrode surface changes adsorption characteristics thereby modifying the sequence of adsorption of different proteins from complex mixtures.

Since direct electron transfer between electrodes and adsorbed redox proteins is not always possible (Hill and Frew 1988) and the construction of electron relay systems is difficult (Bartlett et al. 1987), many of the sensors described use the approved electrochemistry of hydrogen peroxide formed during an enzyme reaction.

Having constructed three types of thin film sensor with platinum coated electrodes we first tried to adsorb GOD either purchased from different suppliers (Sigma and Boehringer) or further purified by chromatofocusing (Fig. 9.3) or size exclusion chromatography (Fig. 9.4).

Trying to desorb GOD from the platinum surface it turned out that protein adsorption was really a chemisorption and therefore under our experimental conditions irreversible (similar results see *Ward et al.* 1990). Neither temperature, saturated NaCl, 3 M KCl, 1 M $CaCl_2$, buffers of different pH could desorb the protein, nor could any significant

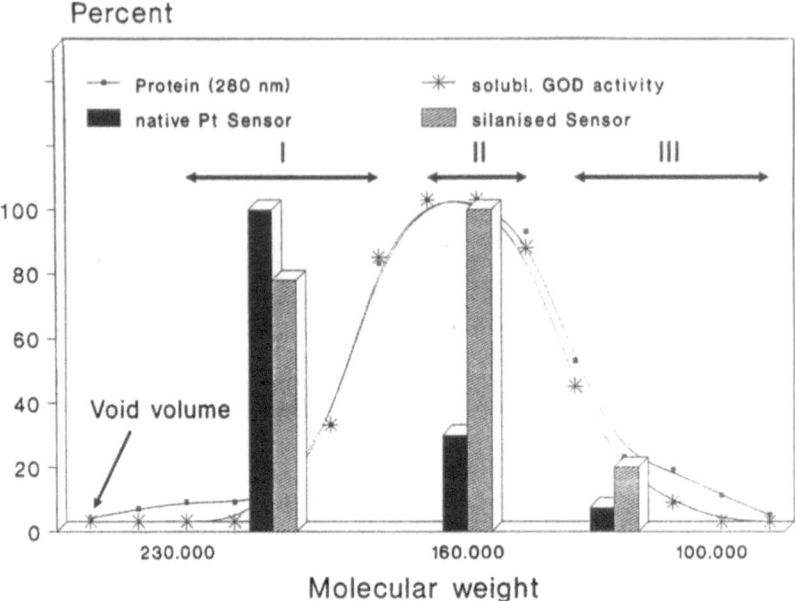

Fig. 9.4. GOD purified by using size exclusion chromatography was adsorbed on an uncoated sensor and on a senor which was coated with aminosilane. The protein containing peak fractions (I - III) were concentrated to 10 mg/ml GOD.

exchange of the adsorbed GOD with high concentrations of soluble protein (BSA, amyloglucosidase, etc) be observed.

Comparing Boehringer GOD grade I , Boehringer GOD grade II, and Sigma GOD type X-S it turned out that neither GOD with the lowest catalase content nor with the highest purification grade or the highest glucose oxidase activity in solution gave sensors with the highest response. Table 9.1 gives a short comparison.

Table 9.1. Glucose sensor response using different types of GOD
CF = eluted peak after chromatofocusing)

type of GOD	relative activity of GOD - preparation	relative response of sensor (4.5 mM glucose)
Boehringer grade I	76%	13%
Boehringer grade II	71%	24%
Sigma grade X-S	55%	17%
Boehringer grade I CF	100%	13%
Boehringer grade II CF	79%	100%
Sigma grade X-S CF peak 1	95%	83%
Sigma grade X-S CF peak 2	78%	45%

Three preparations of GOD having different degrees of purity were obtained for adsorptive and covalent immobilisation by using size exclusion chromatography to purify GOD (Boehringer Grad II), thereby splitting the active peak into fractions (I to III) and concentrating them by ultrafiltration (using a membrane with a cut off of

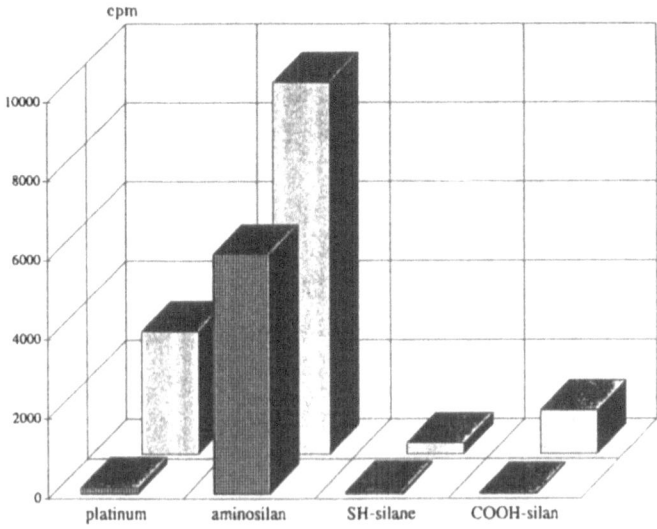

Fig. 9.5. The adsorption of a 20 -mer oligonucleotide (radiolabelled with 32P on its 5 min end) was tested on uncoated platinum, an aminosilane coated sensor, a mercaptosilane coated sensor (SH-silane) and on a mercaptosilane coated sensor which was further derivatised with iodoacetic acid.

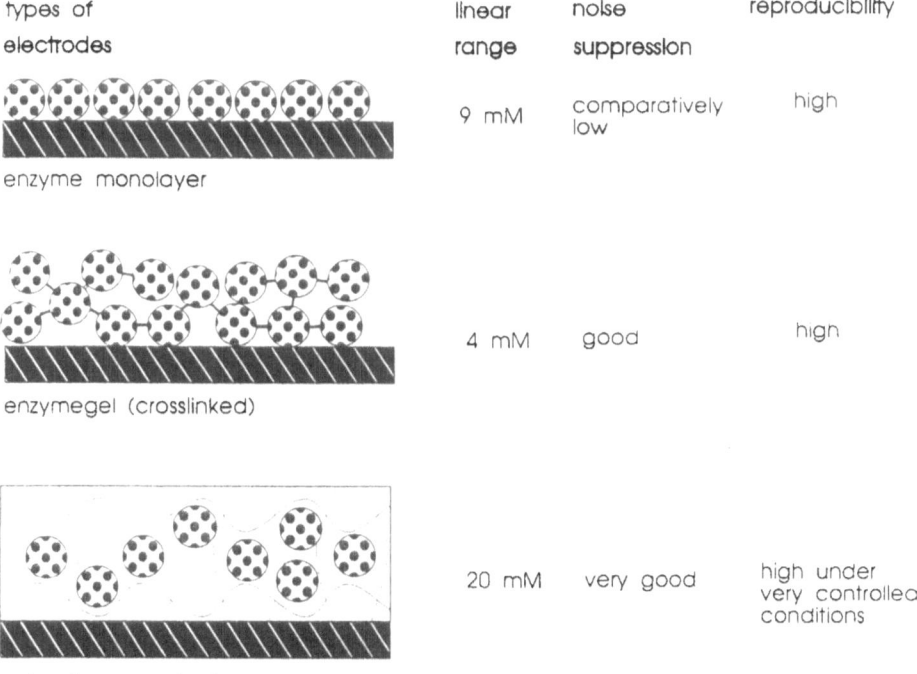

types of electrodes	linear range	noise suppression	reproducibility
enzyme monolayer	9 mM	comparatively low	high
enzymegel (crosslinked)	4 mM	good	high
gel entrapement of enzyme	20 mM	very good	high under very controlled conditions

Fig. 9.6. Different types of electrochemical enzyme biosensor.

30 000 dalton) up to a concentration of 10 mg/ml glucose oxidase. As shown in Figure. 9.4 adsorptive immobilisation on a native platinum surface, having a negative net charge, is highly influenced by impurities being present in low but effective concentrations even from the molecular weight 180 000 down to about 80 000. Using a silanised surface having a positive net charge only impurities with a molecular weight lower than 140 000 will influence GOD adsorption.

Adsorption of Oligonucleotides

Testing the adsorption of DNA (using the ^{32}P-modified 20 bp oligonucleotide : ATTGCGGGTTCTAATCCAGA, Tm = 58°C) on several types of silanised, oxidised and unmodified sensors, it turned out that only a mercaptosilane layer provides a very low adsorption background (Schalkhammer *et al.* 1991).

Further activation of the thiol groups of mercaptosilane coating by nucleophilic substitution with iodoacetic acid and coupling of aspartic acid with carbodiimide did not increase adsorption on the sensor surface (Fig. 9.5).

Covalently Bound Monolayers

Though immobilisation by adsorption is the simplest way to bind enzymes and DNA on metal and carbon electrodes significant leakage (Fig. 9.6a) and protein denaturation cannot be prevented in many cases. Covalent binding turns out to be more feasible. using a number of functional groups which can be activated by well defined organic or inorganic reactions avoiding undefined chemisorption. Nearly all methods well known from other fields of immobilisation can be adapted to the demands of thin layer biosensors. To establish covalent bonding to metal surfaces it is necessary to modify the surface by oxidation or complex formation with thiols or activated nitrogen compounds. Oxidised surfaces can be modified by reacting with silane compounds having pending functional groups (e.g. amino-, thio-, epoxy-, etc). By employing bifunctional reagents, proteins and DNA can be coupled to the surface.(Fig. 9.7)

Techniques

Activation of the Metal Surface

Platinum electrodes were reduced using either 1% sodium dithionite or 5% FeSO₄/6n HCl and oxidised in a solution of 2.5% potassium dichromate in 15% nitric acid for 30 min at 56°C. The electrodes were rinsed with water, followed by acetone, dried and immediately derivatised with silane.

Coupling with 3-amino or Mercaptopropyl-triethoxysilane

The oxidised electrodes were incubated in a 5% aqueous solution of 3-aminosilane pH = 3.5 at 37°C for 30 min. The sililated surface was cleaned with water and ethanol. Using an alternative procedure the oxidised electrodes were incubated in a 7% aqueous

a

b

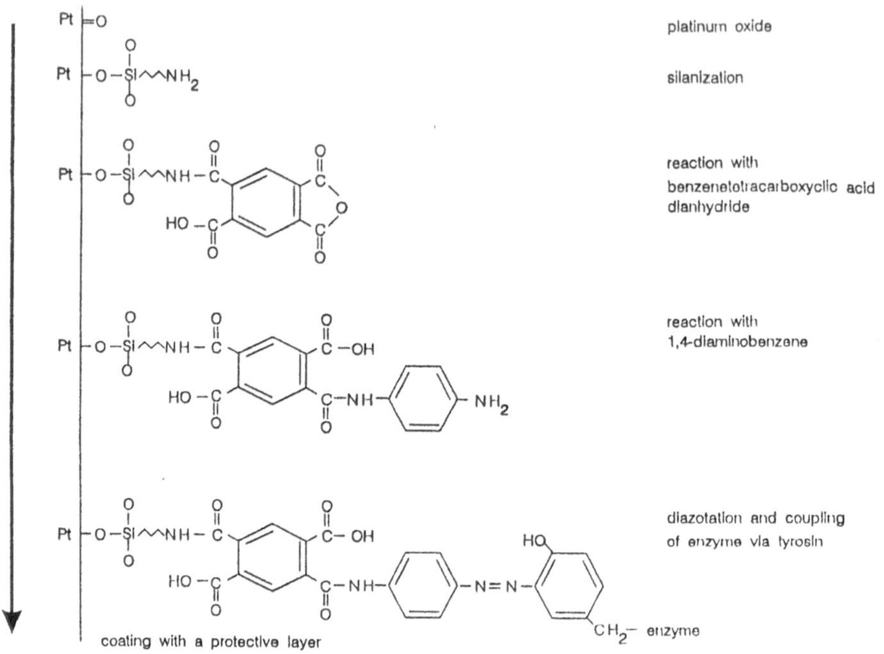

Fig. 9.7. Techniques for covalent immobilisation of enzyme and DNA in biosensor production.

solution (containing 50% acetone) of 3-mercaptosilane pH = 4 at 37°C for 30 min. To enhance silane layer stability the electrodes were further crosslinked by drying them at 110°C for 15 min.

Not only metal electrodes but also conducting glasses can be activated by a similar procedure (Okawa *et al.* 1989).

Activation by p-quinones (Procedure 1)

In addition to p-chloranil also p-halogenanils such as p-fluoranil or p-bromanil and related quinones were used to activate the electrode surface by incubating the silanised sensors with a 1% solution of the reagent for 30 min at 25°C. After activation with quinones the sensors can be stored at 4°C in the dark. To couple e.g. proteins the sensor should be incubated in an aqueous solution of the reagent(10mg/ml) for 2 h (Fig. 9.7a). For further details see Pittner *et al.* 1989, 1990a and 1990b.

Activation by Benzene Tetracarboxylic Acid Dianhydride

The silanised sensors were activated by a 1% solution of benzene tetracarboxlic acid dianhydride (BCTA) in anhydrous THF for 30 min at 25°C (Fig. 9.7b). After washing with anhydrous THF the sensors were immediately dipped into the enzyme solution (10 mg/ml phosphate buffer pH7) and held at 4°C for 2 h.

The BCTA activated sensors can be further derivatised by diazotation (Fig. 9.7c). After reaction with 2% diaminobenzene in THF for 30 min the sensors were washed with 1% acetic acid. and dipped in freshly mixed icecold aqueous solution of 0.25M sodiumnitrite, 1 N in HCl and held for 30 min at 0°C. After washing with ice cold water the enzyme was coupled as described before.

Activation by Iodoacetic Acid and Carbodiimide (Procedure 2)

Mercaptosilane coated electrodes were incubated with iodoacetic acid (1%, pH = 8.0) for 30 min at room temperature. Sensors were rinsed with water and activated with water soluble carbodiimides e.g. by incubating the sensor in an aqueous solution of N-cyclohexyl-N'-[2-(N-methylmorpholino)- ethyl]- carbodiimide- 4-toluene sulfonate (CDI) (60 mg/ml) for 90 min at room temperature. After activation with CDI the sensor was coupled immediately with a protein or an oligonucleotide in aqueous solution (60 min).

Activation by Chloranil, Thioacetic Acid and Carbodiimide (Procedure 3)

To activate the electrode surface the silanised sensors were incubated with a solution of chloranil (1% in toluene) for 30 min at 25°C, washed with toluene and acetone, reacted with thioacetic acid (10% solution buffered to pH = 6.0) for 30 min at room temperature and coupled with carbodiimide as listed above. After activation the sensors should be coupled immediately with e.g. an oligonucleotide in aqueous solution (100 pM/ml, 2 h) .

Activation by Iodoacetic Acid, Carbodiimide, Diamine and Carbodiimide (Procedure 4)

Electrodes were activated with iodoacetic acid and carbodiimide as described, incubated with 0.2 M 1,6-diaminohexane (pH = 6.0, 25°C) for one hour and coupled by incubating in 0.1 M imidazol, 0.1 M water soluble carbodiimide and oligonucleotide (100pM/ml) at 25°C for 12 h.

Techniques for Oligonucleotide Immobilisation

For covalently coupling ss-DNA two strategies can be used :

i   Random coupling using pending (e.g. amino) groups of the oligonucleotide or
ii Selective immobilisation using a terminal (e.g. phosphate) group.

Both strategies were tested and compared. For random immobilisation mercaptosilane coated electrodes were activated by either chloranil (procedure 1) or iodoacetic acid and carbodiimide (procedure 2) or chloranil, thioacetic acid and carbodiimide (procedure 3).

Procedures 2 and 3 lead to a surface with a high affinity for all primary amino groups, procedure 1 prefers linkage to thiols or some highly active amino groups only.

For terminal immobilisation mercaptosilane coated electrodes were activated by iodoacetic acid, carbodiimide, diaminohexane (as a spacer) and carbodiimide (procedure 4) (Ghosh and Musso 1987).

Immobilisation, using chloranil, was able to couple proteins but was not able to couple oligonucleotides. Using procedure 1 and procedure 2 (about 50% less effective than procedure 1) we were able to immobilise DNA efficiently via amino groups. Terminal immobilisation (procedure 4) was not so effective but due to a higher mobility of the immobilised strands of DNA the hybridisation properties are superior (Wolf et al. 1987).

Polymer Supports

Given the well defined structure of the microelectrodes, electrochemically prepared polymer coatings are among the most studied carriers for enzyme coupling. Due to the simple preparation at moderate pH and ion strength (e.g. pH = 6.0, 1 M KCl) especially polypyrrole is suitable for glucose oxidase immobilisation. A number of glucose sensors have already been constructed dealing with the immobilisation of glucose oxidase by copolymerisation with pyrrole and adsorption on or inclusion in pyrrole polymers (e.g. Tien et al. 1988). However, sensors using polypyrrole as an electrochemical active electrode and a mediated electron transfer (by e.g. ferrocene) are independent of oxygen (Lowe and Foulds 1988) but have a low operational stability due to mediator leakage and destruction of the mediator and the polypyrrole matrix by oxygen and hydrogen peroxide.

Considering the limits of enzyme monolayer electrodes, two fundamental advantages of electrodes covered by microporous substituted polypyrrole layers (having no redox activity under working conditions) showed up :

i   significant increase of response per unit area due to the porous surface;
ii permeation control by the polymeric layer for interfering electroactive substances
    (e.g. ascorbate, thiols, etc) resulting in a distinct increase in selectivity.

Moreover the oxygen demand of the sensor is significantly decreased due to the high permeability of the bulk polypyrrole for oxygen and a significantly lower permeability for glucose.

Optimal results for these electrodes could only be obtained through synthesis of modified polypyrrole layers without significant conductivity and redox activity under working conditions (Pittner et al. 1990c and 1991). Thus the electrocatalytic reaction should take place at the metal surface, otherwise high background currents and low selectivity will be obtained.

For the coating of platinum electrodes with polymeric layers new strategies in the synthesis of 1- and 3-substituted pyrroles have been developed to obtain e.g. 1-(carboxyalkyl)-pyrroles, 2-(1-pyrrolo)-acetylglycine, 1-alkylpyrroles, 1-(4-carboxybenzyl)- pyrrole, 1-(4-nitrophenyl) pyrrole, 4-(3-pyrrolo)-4-ketobutyric acid, 3-((keto 4-nitrophenyl) methyl) pyrrole as monomers. The synthesised pyrrole derivatives were tested for their ability to form polymers. By electrochemical oxidation and polymerisation of these monomers in organic solvents new types of polymer coated electrodes were prepared and characterised.

For optimal results homopolymers, heteropolymers and sandwiched types of these polymer layers were prepared. The various layer types led to different properties of the resulting enzyme sensors. Satisfactory homopolymer layers could only be formed with pyrrole derivatives having no bulky side groups directly attached to the pyrrole ring. Also it is vital to increase hydrophobicity of carboxy substituted pyrroles by using long chain derivatives otherwise the polymer films will be soluble in the neutral and slightly alkaline buffered solutions the sensor should be used in.

To overcome the problems due to increased solubility copolymerisation with pyrrole is an efficient way to get aqueous stable and reactive polymer films. In synthesising these heteropolymers it was important to pay attention to the fact that the compound with the more positive polymerisation potential will be incorporated into the polymer to a much lower extent than its concentration in the polymerising solution would normally allow.

When making copolymers of all the above mentioned modified pyrrole monomers with pyrrole it turned out that the most hydrophilic compounds gave the most promising copolymers for sensor preparation. 2-(1-Pyrrolo)-acetylglycine forming only thin water soluble polymer films can be copolymerised with pyrrole to obtain porous films having an optimal enzyme load.

COOH- and nitro-groups which are stable against oxidation under polymerising conditions have proved to be optimal to obtain a modified polypyrrole layer which can be used for covalent coupling of enzymes using water soluble carbodiimides and chloranil as activating reagents (for procedures see covalent immobilisation).

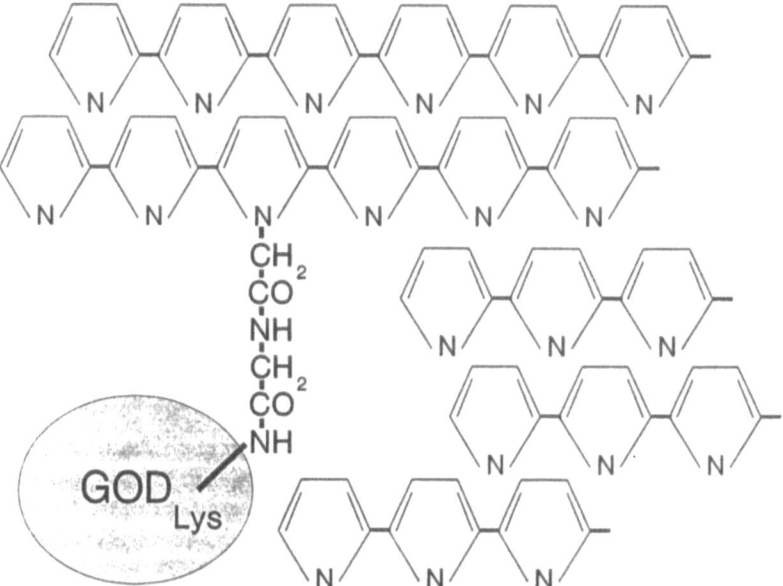

Fig. 9.8. Porous polypyrrole GOD sensor: GOD coupled to a 2-(1-pyrrolo)-acetylglycine polymer using watersoluble carbodiimide.

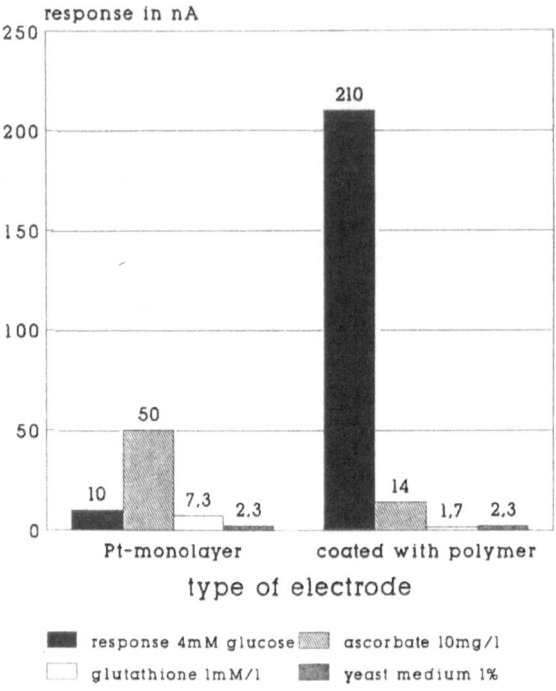

Fig. 9.9. Comparison of a Pt-GOD monolayer and a Pt/ polypyrrole/ GOD electrode in respect to unspecific response of interfering substances.

For the coupling of the enzyme the pore size of the polymer layer must be at least 10 nm - 20 nm. Due to the fractal growth of the polymer on the electrode (Hendrickson *et al.* 1989) during polymerisation slight changes in cycling speed, water content, purity of reagents, etc. would have a great influence on the pore size and inner surface of the polymer film. So it is vital to optimise the polymerisation of each sensor lot and polymerisation mixture.

The activated electrodes were reacted immediately with glucose-oxidase and the glucose sensors thus obtained were stored at 4°C (Fig. 9.8).

As it can be seen in Figure 9.9 , the substituted polypyrrole layer acts as an effective barrier for interfering redox active compounds such as ascorbic acid, bioactive amines, sulfhydryl containing peptides and proteins, etc but having a high permeability for e.g. hydrogenperoxide.

A characterisation of a typical electrode is given below:

i   Porous polypyrrole immobilised glucose oxidase has a 10 - 50 fold greater response/sensor area than monolayer electrodes due to a significant increase in inner surface.

ii The temperature dependence of the signal is 5%/°C and the base current is 1.5 - 3 nA/mm^2.

iii The response time is in the range of a few seconds but increasing if the electrode is covered by a further membrane.

iv The response of polypyrrole coated electrodes is nearly independent of fluctuations in the test solution.

v A high reproducibility is difficult to achieve when immobilising on the internal surface of a polymer layer.

vi The unspecific response of interfering redoxactive substances (e.g. ascorbic acid, phenacetine, etc) can be suppressed by either using the 4-electrode structure and employing differential measurement or using the polypyrrole type of electrodes exhibiting permeation selective properties.

Typical properties of an 2-(1-pyrrolo) - acetylglycine polymer coated glucose sensor will be documented in the following: The change of response as a function of voltage has the same shape as on a native platinum electrode. Under optimum conditions the electrodes working at a potential of 500 mV/(Ag/AgCl) gave a linear response up to 25 mM without any further diffusion limiting membrane. By covering the electrode with a polymer membrane, the linear range for glucose can be extended to more than 60 mM.

Techniques

Cleaning of the Electrode

Platinum thin film electrodes were cleaned by ultrasonication in distilled water, rinsed with acetonitrile and dried carefully under dust free conditions. The electrodes were cycled five times in acetonitrile/2.5% $LiClO_4$ between -500 mV and 1800 mV (100 mV/s) versus Ag/AgCl using a 3 electrode configuration of an electrochemical cell. A potentiostat interfaced by a 14 bit A/D D/A converter to an AT personal computer was employed for generating the required voltages.

Coating with Polymerised Substituted Homopolypyrroles

Table 9.2. Polymerisation of substitued homopolypyrroles

Compound	U (Ag/AgCl)*	polymer-redox peaks[+]
A 1-(2-Carboxyethyl) - pyrrole	soluble polymer	
B 1-(3-Carboxypropyl) - pyrrole	soluble polymer	
C 1-(5-Carboxypentyl) - pyrrole	1200 mV	350-830 mV
D 1-(10-Carboxydecyl) - pyrrole	1200 mV	350-830 mV
E 2-(1-Pyrroleo) - acetylglycine	1200 mV	550-800 mV
F 1-Dodecylpyrrole	1200 mV	550-780 mv
G 1-(4-Carboxybenzyl)- pyrrole	1200 mV	450-750 mV
H 1-(4-Nitrophenyl) pyrrole	1400 mV	730-770 mV
I 1-(4-Carboxyphenyl) pyrrole	only copolymer	
J 4-(3-Pyrroleo)-4-ketobutyric acid	1300 - 2000 mV	
K 3-((Keto 4-nitrophenyl) methyl) pyrrole	1800 mV	850-950 mV

* to obtain stable polymer films
[+] in Acetonitrile/$LiClO_4$

The compounds above (see Table 9.2) were used as monomers for coating the platinum electrodes with thin polymeric layers (Pittner *et al.* 1990c and 1991).

A 0.5% solution of the above mentioned compounds in acetonitrile containing 2.5% of $LiClO_4$, NR_4BF_4 or NR_4PF_6 (R = Me,Et,Bu) was dried over Na_2SO_4 or $CaCl_2$ for several hours. The clear solution was transferred to a glass cell and bubbled with argon for 10 min. The electrode immersed in the de-aerated solution was now cycled 10 - 30 times between -300 mV and 1800mV the potential listed above (100 mV/s). If the substituted pyrrole was polymerisable the formation of a thin, brown polymeric layer was observed. Most of these polymers (see Table 9.3) showing redoxactivity in acetonitrile/$LiClO_4$ were redoxinactive in aqueous solutions at pH = 7.0.

Table 9.3. Stability, solubility and permeability of polymer films

Compound -->	A	B	C	D	E	F	G	H	I	J	K
stability of the polymer film in aqueous buffer pH=7	0	0	+	+	-	+	+	+	+	-	+
redoxactivity of polymer film in aqueous buffer pH=7	0	0	+	+	-	+	-	-	-	-	-
permeability for hydrogenperoxide	0	0	+-	+-	0	-	+-	+-	-	-	0

For most sensor applications it is necessary to make a copolymer having a high content of unsubstituted polypyrrole in order to obtain thick and stable substituted polypyrrole films. Since the polymerisation speed of the substituted pyrroles is significantly lower than that of unsubstituted pyrrole it is, nevertheless, necessary to have a 10:1 excess of the substituted monomer.

The electrolyte of the above mentioned composition was complemented with additional 0.05% pyrrole. The electrode was cycled up to 10 times between -300 mV and +1400 mV (100 mV/s). A black polymer layer of varying thickness was obtained. For coupling of the enzymes a variety of methods was employed to activate the terminal carboxylic acid groups. The use of water soluble carbodiimides may be recommended for best results.

Polypyrrole coated electrodes having COOH-groups were incubated with a saturated solution of N-cyclohexyl-N'-[2-(N-methylmorpholino)-ethyl]-carbodiimide-4-toluenesulfonate for 30 min at 25°C without shaking. The electrodes were rinsed several times with water and immediately reacted with a 5 mg/ml enzyme solution in phosphate buffer pH = 7.0 ; 0.1 M for two hours.

Substituted polypyrrole films having nitro-groups were reduced using a solution of 1% $SnCl_2$ in 10% HCl for 30 min forming a film with pending amino- groups. After excessive rinsing with diluted HCl, distilled water and ethanol, these electrodes were dried. Now 1,4 - arenequinones were employed to couple the enzyme. The sensors thus obtained were rinsed with phosphate buffer and water and stored at 4°C.

Membranes

As can be seen in the literature membrane sensor technology is a well established field in biosensor production. The glucose analyser from Yellow Springs Instrument was one of the first analysers employing cellulose acetate multi membrane technology for glucose monitoring. High oxidase loading and low interfering currents are among the main advantages. However, the macroscopic construction of the sensor assembly, an increased response time and its complicated use for diabetes monitoring limits the application of this sensor technology. Moreover, only thick and thin film technology is able to produce the high numbers required. Whereas thick film technology is able to produce simple and low integrated sensors, membrane technology is a developing field also for complicated and highly integrated sensors using thin film technology.

To reach this aim dip-, spray -and spin coatings sometimes combined with photostructuring techniques are the most practical.

Using oxidase based biosensors three types of membrane are recommended :

i   a hydrogen peroxide permeable hyper filtration membrane with a low cut off on
     top of the metal electrode to suppress the action of interfering substances
ii   an enzyme containing membrane (if necessary)
iii a top membrane for biocompatibility

The use of glutaraldehyde for the preparation of enzyme membranes has already been described 20 years ago (Fig. 9.6b). Most of these membranes are composed of the desired oxidase, mixed with carrier proteins like BSA and crosslinked with glutaraldehyde (Schmid *et al.* 1991). The gels are not well defined and subjected to sometimes irreversible shrinking or swelling. An advanced technique is presented using a new type of glutaraldehyde - diamine (amine) based polymer. At neutral and slightly alkaline pH amines reacting with glutaraldehyde do not form single enamine bonds but lead to linear high molecular weight polymers. Employing the reaction procedure described above and using mixtures of amines and diamine compounds results in stable crosslinked polymers of the desired properties. Using 1,6 - diaminohexane as reacting and crosslinking amine (1 M) and glutaraldehyde (2.5 M - 3 M) at an pH of 7 - 9 a hydrophilic gel will result. Two fundamental procedures are given below.

Spinning techniques are the most practical for forming thin and ultrathin membranes (0.01 μm - 10 μm). Many preformed polymers can be dissolved to obtain highly viscous fluids which can be used directly to spin membranes onto the surface of sensor substrates. Water soluble polymers can be further crosslinked by a variety of multifunctional reagents (photocrosslinkers, epichlorhydrine, isocyanates, etc) to obtain hydrophilic, but water insoluble networks. Being embedded in a hydrophilic network to obtain polyol containing cages many enzymes show increased stability. A good example of this technique with lactate oxidase from *Pediococcus* sp. a FMN containing enzyme is given in the section on techniques.

Techniques

Coating with Diamine/Glutaraldehyde Membrane

An aqueous solution of 1,6-diaminohexane (15%, adjusted to pH = 7.5 with acetic acid) 5 μl , 0.1 M phosphate buffer (0.1 M, pH = 7.0) 5 μl and glutaraldehyde (25%) 10 μl were mixed and immediately used for coating the sensor by the above mentioned techniques. To prevent polymer from peeling off, the sensors should by dipped into Triton X-100 (0.1%) before coating with the polymer layer. After crosslinking for 60 min at room temperature an oxidase can be coupled to the membrane by simply dipping the coated sensor into a buffered solution of the enzyme 0.1 mg/ml - 10 mg/ml for 60 min. To achieve a high stability of the immobilised enzyme under dry storage conditions a further layer of protein e.g. hemoglobin should be immobilised on top of the enzyme layer by dipping the sensor into buffered glutaraldehyde (5%, pH = 7.0) for 30 min at 25°C followed by washing and incubating with hemoglobin (1%, pH = 7.0 , 30 min, 25°C).

Coating with Diamine / Glutaraldehyde / GOD Membrane

An aqueous solution of 1,6-diaminohexane (15%, adjusted to pH = 7.5 with acetic acid) 5 μl , a buffered solution of an oxidase (0.1 mg/ml - 10 mg/ml, pH = 7.0) 5 μl, glutaraldehyde (25%) 10 μl were mixed and immediately used for coating the sensor by the above mentioned techniques. To prevent polymer from peeling off the sensors, should by dipped into Triton X-100 0.1% before coating with the polymer layer. To achieve a high stability of the immobilised enzyme under dry storage conditions a further layer of protein e.g. hemoglobin should be immobilised in the gel by dipping the sensor into buffered glutaraldehyde (5%, pH = 7.0) for 30 min at 25°C followed by washing and incubating with hemoglobin (1%, pH = 7.0 , 30 min, 25°C).

Coating with Poly - HEMA Biocompatibility Membrane

Glycerol 500 μl, water 500 μl, 2-OH-ethylmethacrylate (HEMA) 100 μl, aqueous solution of ammoniumperoxodisulfate (10%) 10 μl and tetraethylmethylenediamine 2 μl were mixed and immediately used for spin - or dip coating.of the electrode. The polymerisation takes about 45 min. Before using for measurement the electrodes should be washed and pretreated with buffer for 30 min to swell the polymer layer. To obtain optimal polymer layer stability, the coated electrodes should be stored in a vapor saturated atmosphere at 4°C.

For the use of poly HEMA as enzyme carrying membrane see Pickup *et al.* 1991.

Lactate Oxidase Sensor using Poly(vinylalcohol) Membranes

The electrode is covered by a poly(vinylalcohol) (PVA) membrane by spinning 20 μl of an aqueous solution (3%) of PVA for 30 s at 5000 rpm. An enzyme layer is spun on top of the polymer layer using 20 μl of lactate oxidase solution containing 10 units of the enzyme in bidistilled water. A further PVA membrane (prepared similar to the first one) is used to protect the enzyme layer. To obtain a water insoluble network, the PVA toplayer is crosslinked by dipping the electrode in a solution of diphenylmethane diisocyanate (12%) in tetrahydrofuran for 1-5 min, washing immediately with tetrahydrofuran (20 s) and immersing the electrode in phosphate buffer pH = 7.0.

A similar procedure for platinised carbon electrodes is given by Heineman et al. 1991.

Discussion

In principle all techniques given in this chapter are very well suited to constructing thin film biosensors. To take the advantage of thin film techniques so as to be able to produce sensor batches in high lot numbers it is recommended to use dip - and spinning procedures avoiding e.g. electrochemical polymer formation. For practical applications, a high signal to noise ratio can only be obtained using selective hydrogen peroxide permeable membranes (Fig. 9.10), differential measurement or mediators on carbon coated surfaces. Using chemically modified monolayer electrodes it is vital to pay

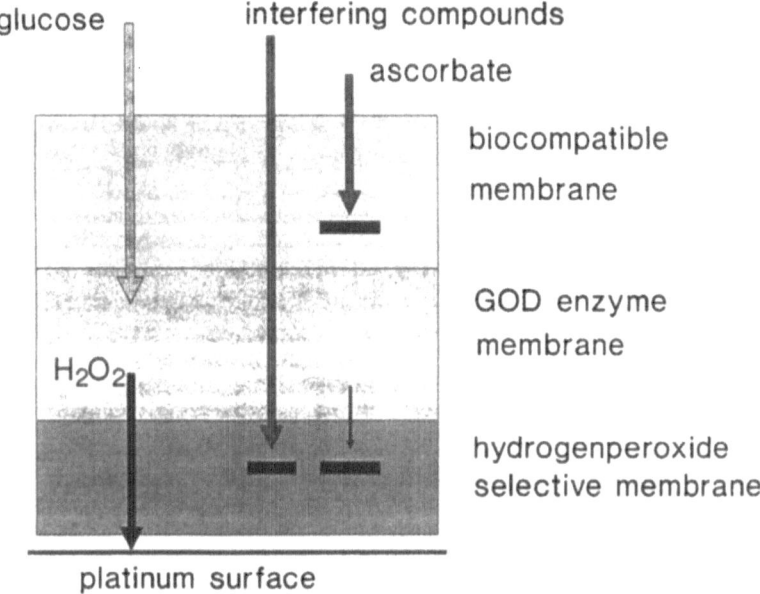

Fig.9.10. Interfering reactions on biosensor surfaces.

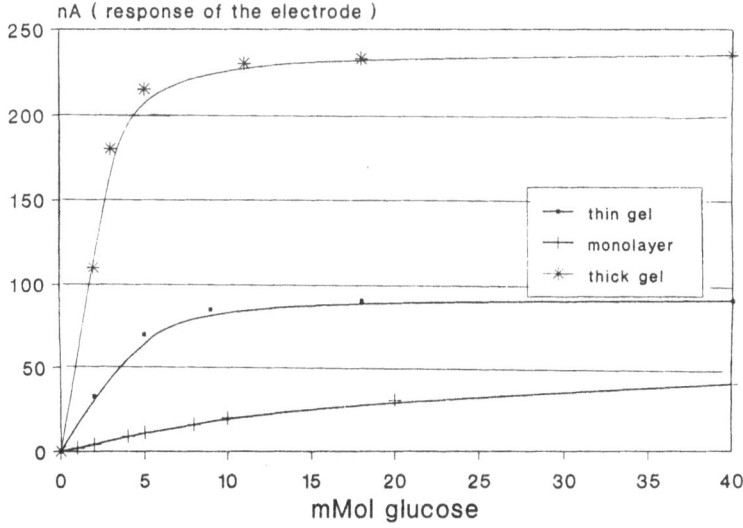

Fig. 9.11. Calibration curves of different sensor types: thin gel: GOD crosslinked with GDA; monolayer: chloranil or benzene tetracarbonic acid dianhydride coupled GOD; thick gel: GOD crosslinked with diaminohexane/ GDA.

attention to the modified and in most cases suppressed electrochemistry due to an oxidised or poisoned metal surface. To get rid of impurities at the platinum surface argon etching can be used to sputter off about 10 nm of contaminated metal.

A high enzyme concentration at the electrode surface will cause a significant decline in the linear range of the biosensor due to oxygen consumption. It is therefore recommended to use only monolayers or, if not feasible, highly diluted enzyme in the membrane. To increase the linear range the diffusion limiting layer should be as hydrophobic as possible because of the higher solubility of oxygen in hydrophobic media.

Comparing all the advantages and disadvantages of the various sensor types (Fig. 9.11; Fig. 9.6) it seems that adsorption and thin membrane technology is the most simple way to obtain oxidase based enzyme sensors. However covalently bound monolayers may display a variety of advantages for the construction of immuno - and DNA-biosensors. Sensors coated with electrochemically prepared polymers are suitable for special applications not requiring high reproducibility at high numbers.

Acknowledgement

Part of this work was supported by a grant of the Fonds zur Förderung der wissenschaftlichen Forschung No. S-4908 and a grant of the Bundesministerium für Wissenschaft und Forschung GZ-49.688/8 - 27a/90.

References

Bartlett PN, Whitaker RG, Green MJ, Frew J (1987) Covalent binding of electron relays to glucose oxidase. J Chem Soc Chem Commun 1603-1604

Bergveld P (1991) A critical evaluation of direct electrical protein detection methods. Biosensors and Bioelectronics 6:55-72

Ghosh SS, Musso GF (1987) Covalent attachment of oligonucleotides to solid supports. Nuc Acids Res 15:4485-4502

Hendrickson WA, Yang R, Dalsin KM, Evans DF, Christensen L (1989) Scanning tunneling microscopic imaging of electropolymerised, doped polypyrrole. Visual evidence of semicrystalline and helical nascent polymer growth. J Phys Chem 93:511-512

Heineman WR, Hajizadeh KH, Halsall HB (1991) Immobilisation of lactate oxidase in a poly(vinyl alcohol) matrix on platinised graphite electrodes by chemical cross-linking with isocyanate. Talanta 38(1):37-47

Hill AO, Frew JE (1988) Direct and indirect electron transfer between electrodes and redox proteins. Eur J Biochem 172:261-269

Lowe CR, Foulds NC (1988) Immobilisation of glucose oxidase in ferrocene-modified pyrrole polymers. Anal Chem 60:2473-2478

Okawa Y, Tsuzuki H, Yoshida S, Watanabe T (1989) Glucose sensor carrying monomolecular layer of glucose oxidase covalently bound to tin (IV) oxide electrode. Anal Sci 5:507-512

Pickup JC, Shaw GW, Claremont DJ (1991) In vitro testing of a simply constructed, highly stable glucose sensor suitable for inplantation in diabetic patients. Biosensors and Bioelectronics 6(5):401-406

Pittner F, Schalkhammer T, Urban G, Mann-Buxbaum E (1989) Verfahren zur Immobilisierung von Proteinen, Peptiden, Liganden (Coenzymen, Antigenen Effektoren,..) an einem Trägermaterial, welches Hydroxy-, Mercapto- und Aminogruppen besitzt. Patent A-786/89 and A-1119/89 applied in 12 countries 1990

Pittner F, Schalkhammer T, Urban G, Mann-Buxbaum E (1990a) New microminiaturised glucose sensors using covalent immobilisation techniques. Sensors and Actuators B1:518-522

Pittner F, Schalkhammer T, Urban G, Mann-Buxbaum E (1990b) Biosensors on thin film metals and polymer coated electrodes. Fresenius J Anal Chem 337(1):107

Pittner F, Schalkhammer T, Urban G, Mann-Buxbaum E (1990c) Electrochemical biosensors on thin film metals and conducting polymers. J Chromatography 510:355-366

Pittner F, Schalkhammer T, Urban G, Mann-Buxbaum E (1991) Electrochemical glucose sensors on permselective nonconducting substituted pyrrole polymers. Sensors and Actuators B4:273-281

Schalkhammer T, Pittner F, Hartig A, Moser I (1991) Surface modification of platinum based electrochemical thin film electrodes for DNA biosensors. Microsystem Technologies Berlin 76-81

Schmid RD, Bilitewski U, Rüger P (1991) Glucose biosensors based on thick film technology. Biosensors and Bioelectronics 6(4):369-373

Tien HT, Kotowski J, Janas T (1988) Immobilisation of glucose oxidase on a polypyrrole-lecithin bilayer lipid membrane. Bioelectrochem. Bioenerg. 19:277-282

Wahlgren M, Arnebrant T (1991) Protein adsorption to solid surfaces.Trends Biotechnol 9:201-208

Ward MD, Ebersole RC, Miller JA, Moran JR (1990) Spontaneously formed functionally active avidin monolayers on metal surfaces: A strategy for immobilising biological reagents and design of piezoelectric biosensors. J Am Chem Soc 112:3239-3241

Wingard LB, Miyawaki O (1984) Electrochemical and enzymatic activity of flavin adenine dinucleotide and glucose oxidase immobilised by adsorption on carbon. Biotech Bioeng 26:1364-1371

Wolf SF, Haines L, Fisch J, Kremsky JN, Dougherty JP, Jacobs K (1987) Rapid hybridisation kinetics of DNA attached to submicron latex particles. Nuc Acids Res 15(7):2911-2926

Chapter 10

Two-Dimensional (Glyco)protein Crystals as Patterning Elements and Immobilisation Matrices for the Development of Biosensors

D. Pum, M. Sára, and U.B. Sleytr

Introduction

At the present time there is a growing interest in the immobilisation of biologically active macromolecules for the fabrication of a new generation of bioanalytical sensors (Turner *et al.* 1988). It is particulary the combination of the functionality of the immobilised molecules with the physical, electrical, optical or chemical properties of the carrier which is going to revolutionise the analytical sciences. But these developments are not only restricted to the biosensor field. Quite a few other scientific disciplines like microelectronics, microoptics or micromachining are working on a controlled fabrication of biologically active nanometer structures. This multidisciplinary field between physics, biochemistry and molecular biology is now often referred to as nanometer technology and is devoted to the development of molecular machines in the nanometer range (Carter 1983, Douglas and Clark 1986, Andrade 1987, Eigler and Schweizer 1990, Eigler *et al.* 1991, Quate 1991). The controlled immobilisation of functional molecules such as enzymes or antibodies allows the fabrication of submicron devices with a completely new spectrum of useful properties. Whether the macromolecules act as passive or active elements depends on the application and on the design of the device. For example, molecules which change the surface charge or hydrophobicity of a membrane or a solid support are passive elements since they do not generate a physically exploitable signal. The sensing layer in a biosensor which is usually made of enzymes, antibodies or other biologically active components plays an active role because it acts as a molecular machine in the nanometer range. In all aspects of nanometer fabricated devices, the arrangement and availability of functional groups determine the binding capacity of the immobilisation matrix and therefore the usability of the device.

The preparation of a support where functional groups are rather homogeneously distributed over the entire surface with spacings of only a few nanometers may be achieved in two ways. The classical approach lies in the downscaling of existing technologies into the submicron scale and the optimisation of the immobilisation methods in such a way that a rather dense packing of functional molecules becomes possible. Microlithographic techniques are commonly used to deposit metal structures on insulating materials such as glass or silicon. Functional groups may be introduced on such surfaces by conventional silanisation methods. Although the coating of metal and glass surfaces has been optimised over the years a dense homogeneous packing of functional molecules is usually not possible. An alternative approach in preparing patterning elements in the nanometer range has been developed at the Center for Ultrastructure Research in Vienna (Sleytr et al. 1988a, Pum et al. 1989b). The work on the structure, chemistry, assembly and function of 2-D crystalline bacterial surface layers, termed S-layers (Sleytr et al. 1988a) led to further investigations on the application potential of these lattices as immobilisation supports in the development of biosensors and in the nanometer technologies. S-layers are periodic macromolecular structures with a high binding capacity in comparison to conventional immobilisation matrices. Contrary to carriers where the location, local density and orientation of functional groups and the porosity and pore size are only known approximately, with crystalline layers the properties of a single constituent unit are replicated with the periodicity of the lattice and thus define the characteristics of the whole two-dimensional array. According to this principle, we understand by the expression "controlled immobilisation" the process of binding molecules onto a crystal lattice at geometrically and physicochemically well defined locations.

Chemical and Structural Principles of Crystalline Bacterial Surface Layers (S-layers)

Two-dimensional crystalline surface layers have been observed as the outermost cell envelope component in many strains of walled eubacteria and represent an almost universal feature in archaebacterial cell envelopes (Fig. 10.1) (for reviews see Beveridge 1981, Smit 1986, Koval 1988, Sleytr and Messner 1983, Sleytr and Messner 1988, Sleytr et al. 1988a, Messner and Sleytr 1992).

S-layers cover the bacterial cell completely and are organised in the form of oblique, square or hexagonal lattices (Fig. 10.2). Depending on the lattice type, one morphological unit is composed of two, four or six identical subunits which are also called protomeric units. The centre-to-centre spacing of the morphological units can vary between 3 nm and 32 nm while the thickness of S-layers is usually in the range of 5 nm to 10 nm. Due to their structure, crystalline S-layers have pores with well defined morphology and size (Fig. 10.2). Most of the presently known S-layers are composed of a single protein species with molecular weights ranging from 40 000 Da to 220 000 Da. Comparison of amino acid analysis and genetic studies on S-layers from a broad spectrum of bacteria has shown that the crystalline arrays are usually composed of weakly acidic proteins, the content of hydrophobic amino acids is generally high and the cysteine or methionine content low. (Sleytr and Messner 1983). With S-layers it was shown for the first time, that eubacteria and archaebacteria are capable of

Fig. 10.1. Electron micrograph of a freeze-etched cell surface of *Clostridium thermohydrosulfuricum* L111-69. The hexagonal array of glycoprotein subunits (S-layer) completely covers the cell surface. Bar, 100 nm.

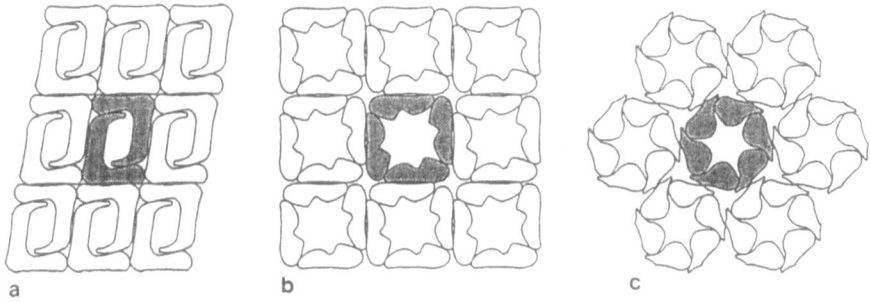

Fig. 10.2. Schematic representation of the major types of S-layer lattices observed on bacteria. **a** Oblique, **b** square and **c** hexagonal lattice symmetry. The unit cells which are the building blocks of the two-dimensional crystalline arrays are formed by two, four or six identical subunits (shaded areas).

glycosylating proteins (Küpcü *et al.* 1984, Messner and Sleytr 1988a, Messner and Sleytr 1988b, Messner and Sleytr 1991). All glycan chains studied are polymers of linear or branched repeating sequencies of two to six monosaccharide units which include a wide range of hexoses, deoxy or amino sugars, uronic acids or even sulphate or phosphate residues as constituents (for reviews see Messner and Sleytr 1991).

Isolation and Purification

Most techniques for isolation and purification of S-layers involve the mechanical disintegration of cells and subsequent differential centrifugation to separate the cell wall

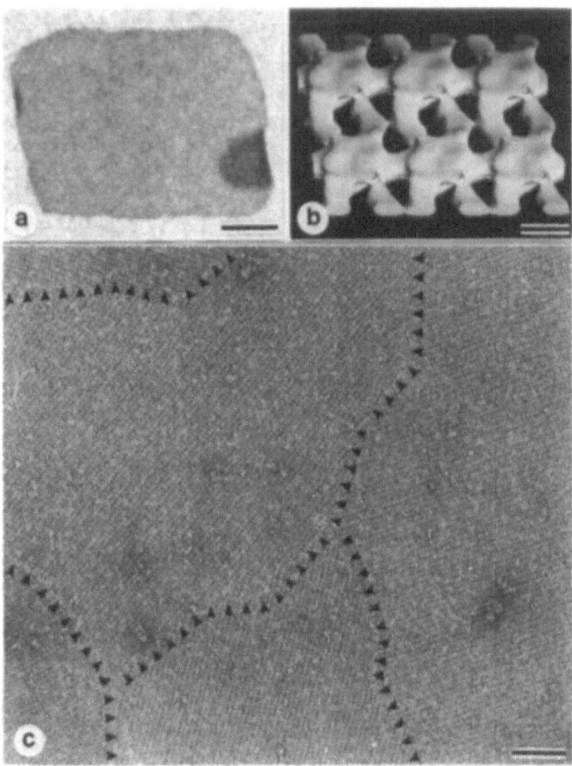

Fig. 10.3. a Negatively stained self assembly product (flat sheet) obtained upon recrystallisation of S-layer subunits isolated from *Bacillus stearothermophilus* NRS 2004/3a. Bar, 200 nm. **b** A 3-D model of the protein mass distribution of the oblique S-layer lattice of *B.coagulans* E38-66. Bar, 5 nm. **c** Monocrystalline patches of the S-layer protein from *B.coagulans* E38-66 obtained by recrystallisation of isolated subunits on a positively charged carbon layer. The crystal boundaries are marked by arrows. Bar, 100 nm.

fragments. A complete disruption of S-layer fragments into their constituent subunits can generally be achieved by a high concentration of H-bond breaking agents (e.g. urea, guanidine hydrochloride) or by changing the pH-value of the medium. Results from different disruption procedures led to the conclusion that:

i   S-layer proteins are not covalently linked to each other;

ii   differing combinations of weak bonds (hydrophobic, ionic and hydrogen bonds) maintain the regular arrangement; and

iii bonds holding the S-layer subunits together are stronger than those binding the lattice to the underlying cell envelope structure (for reviews see Sleytr and Messner 1983, Koval and Murray 1984, Sleytr and Messner 1988).

Self Assembly of S-layer Proteins in Suspension

Frequently, the isolated S-layer subunits maintain the ability to assemble *in vitro* into lattices identical to those observed on intact cells upon removal of the disrupting agent used in the isolation procedure. Studies on this self assembly process in the absence of a template have shown that the initial phase is determined by a rapid nucleation into oligomeric precursors consisting of several unit cells which, in a following much slower process, aggregate into larger crystalline arrays (Jaenicke *et al.* 1985). These may have the form of flat sheets, open ended cylinders, or closed vesicles up to a size of about 10 μm -15 μm (Fig. 10.3.a). The different assembly routes are predetermined by the morphology and bonding properties of the subunits (Sleytr 1978, Sleytr and Messner 1989). With some protomeric units, different assembly routes can be obtained by changing the assembly conditions (pH, temperature, ionic strength, presence or absence of bivalent cations). In addition to monolayers, double layer assembly products are often found. In the case of double layers the two constituent monolayers are bound to each other either with their inner or their outer faces (Messner *et al.* 1986b, Pum *et al.* 1989a). S-layer self assembly products are frequently the basis for high resolution image reconstruction procedures.

Recrystallisation of S-layer Proteins on Solid Supports

S-layer fragments or isolated S-layer subunits from numerous organisms have shown the ability to recrystalise on solid supports such as glass, silicon, mica, carbon or synthetic polymers (Messner *et al.* 1986b, Pum *et al.* 1989a). The environmental conditions are frequently the same as the ones used to induce self-assembly in the absence of supporting layers as described above. As a precondition for many S-layer assembly systems, the supports had to be either treated with alcian blue or poly-1-lysine in order to obtain a positively charged surface. Silanisation with different compounds can be used to produce either positively or negatively charged surfaces on glass or silicon. Depending on the surface properties of the supporting layer and on the S-layer protein the subunits can bind either with their inner or outer face. Chemical modification experiments revealed that the carboxyl and amino groups on the outer S-layer surface are involved in direct electrostatic interactions whereas on the inner face free carboxyl groups are present (Sára and Sleytr 1987a). An example of a reconstituted oblique S-layer lattice on a poly-1-lysine coated carbon film is shown in Fig. 10.3c. The numerous crystal boundaries indicate that the crystal growth is initiated at several distant nucleation points. These can be randomly oriented monomers, oligomers or small crystallites. Depending on the surface properties the individual (mono)crystalline patches adsorb either with their inner or their outer face and grow isotropically in all directions until the front end of the neighbouring crystalline areas is reached. The crystal boundaries between adjacent patches are seen as line imperfections in the crystalline array (Fig. 10.3c).

Ultrastructure

High resolution electron microscopy in combination with digital image processing methods have shown a considerable asymmetry in the topography of the outer and inner

S-layer face. Electron micrographs of differently tilted views of negatively stained or frozen hydrated specimens are processed and combined in the computer (Amos *et al.* 1982). A model of the 3-D protein mass distribution is finally obtained and may be displayed using computer graphics methods (Fig. 10.3b). In general, the outer face of the crystalline array is commonly smooth while the inner face is more corrugated (Baumeister and Engelhardt 1987; Hovmöller *et al.* 1988).

Permeability Properties

S-layers have pores which are identical in size and morphology due to the regular arrangement of the constituent subunits (Sára and Sleytr 1987b). Since they are generated by identical protomeric units they have also identical physicochemical properties. Information on either the mass distribution in S-layer lattices obtained by high resolution electron microscopical procedures or on the "functional pore size" derived from permeability studies revealed that S-layers work in the ultrafiltration range. Depending on the selected strain or organism, the scale of pore sizes ranges from approximately 2 nm - 12 nm. The porosity of S-layers as revealed from computer image reconstructions of electron micrographs is in the range of 30% - 70%. S-layers are used for a completely new type of ultrafiltration membrane (Sára and Sleytr 1987c, Sára and Sleytr 1988, see also Sára *et al.* this volume). For the production of S-layer ultrafiltration membranes (SUM), isolated S-layer fragments are deposited on standard microfiltration membranes and crosslinked with glutaraldehyde which introduces covalent linkages between and within the amino groups of the protein of the individual S-layer fragments.

Binding of Macromolecules to S-layers

Since S-layers as two-dimensional crystalline arrays are composed of identical (glyco)protein subunits it is evident that functional groups such as carboxyl, amino, or hydroxyl groups on the polypeptide chains and/or the carbohydrate moieties of each protomer must be located in an identical position and orientation. Due to this regular arrangement of functional groups S-layers can be considered as a completely new type of matrix for the immobilisation of macromolecules (Sára *et al.* 1988, Sára and Sleytr 1989, Messner *et al.* 1991, see also Sára *et al.* this volume). In the following examples the immobilisation of macromolecules to S-layers through ionic and covalent bonds are described.

Immobilisation of Macromolecules through Ionic Bonds

The distribution of net negatively charged areas on S-layers can be visualised by electron microscopical methods after labelling with charged marker molecules such as polycationised ferritin (PCF). At neutral pH-conditions PCF is a strongly positively charged molecule which can be used due to its size of 12 nm and its strong electron scattering iron containing core region both as a topographical marker and as a label in thin sections. For example, the regular arrangement of free carboxyl groups on the

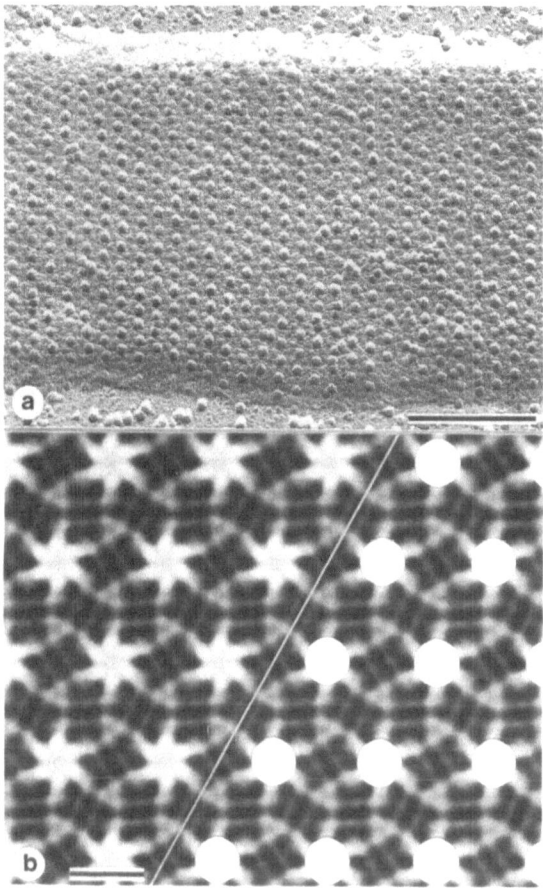

Fig. 10.4. a Freeze-dried and shadowed preparation of the S-layer of the archaebacterium *Thermoproteus tenax* labelled with polycationised ferritin (PCF). The positively charged PCF molecules (diameter = 12 nm) bind in a regular fashion to the negatively charged areas of the hexagonal S-layer lattice. Bar, 200 nm. b Protein mass distribution of the S-layer obtained by computer image reconstruction. PCF molecules which are represented schematically by white discs bind to the protein domains in the center of each morphological unit. Bar, 20 nm.

hexagonally ordered S-layer lattice from *Thermoproteus tenax* could be clearly demonstrated by labelling with PCF (Fig. 10.4) (Messner *et al.* 1986a).

As mentioned above the outer face of the S-layer lattice from the Bacillaceae investigated so far reveals a charge neutral surface due to the presence of an equal number of amino and carboxyl groups (Fig. 10.5a,b) (Sára *et al.* 1989). In order to generate a surface with densely arranged free carboxyl groups the S-layer protein was crosslinked with glutaraldehyde which reacted with the free amino groups. The excess of free carboxyl groups could then be visualised by labelling with PCF which was bound as a monolayer in a dense packing order (Fig. 10.5c,d).

With glycosylated S-layers not only the protein but also the carbohydrate chains which are exposed on the outer face are available for the immobilisation of foreign

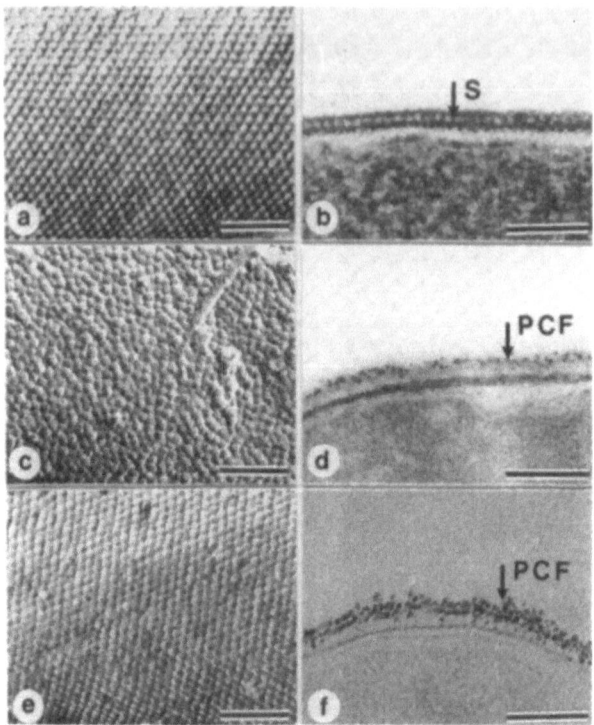

Fig. 10.5. Electron micrographs of freeze-etched preparations (**a,c,e**) and corresponding thin sections (**b,d,f**) of *Clostridium thermohydrosulfuricum* L111-69 labeled with polycationised ferritin (PCF). In thin section preparations the PCF molecules appear as dark dots. a The native S-layer shows hexagonal symmetry. Due to its charge neutral surface no PCF molecules are bound (**a,b**). (**c,d**) After blocking the aminogroups of the S-layer protein by crosslinking with glutaraldehyde, the PCF molecules are bound electrostatically to the free carboxyl groups. (**e,f**) The exposed carbohydrate chains of the S-layer glycoprotein subunits could be labeled with PCF after conversion of the hydroxyl groups into carboxyl groups by succinylation. In freeze etched preparations (**e**) the immobilised PCF molecules reveal a hexagonal packing order. Thin sections (**f**) show in average two to three immobilised ferritin molecules per chain. Bars, 100 nm.

molecules. (Sára *et al.* 1989, Messner and Sleytr 1991). For example, the hydroxyl groups of the glycan chains could be converted into carboxyl groups by succinylation which led to a strongly negatively charged S-layer surface. The presence of the free carboxyl groups along the carbohydrate residues could be demonstrated by labelling with PCF (Fig. 10.5e,f). As shown by ultrathin sectioning, up to four PCF molecules could be bound in line along the exposed carbohydrate chains (Fig. 10.5f). Freeze-etching preparations clearly demonstrated that at least the outermost layer of PCF molecules was bound in a hexagonally ordered fashion resembling the maximum dense packing order (Fig. 10.5e).

Comparative electron microscopical studies on native and glutaraldehyde treated S-layers from different organisms demonstrate that, depending on the type of S-layer lattice and the chemical modifications, different binding patterns of foreign molecules in regular or non regular arrangement could be obtained (Fig. 10.6).

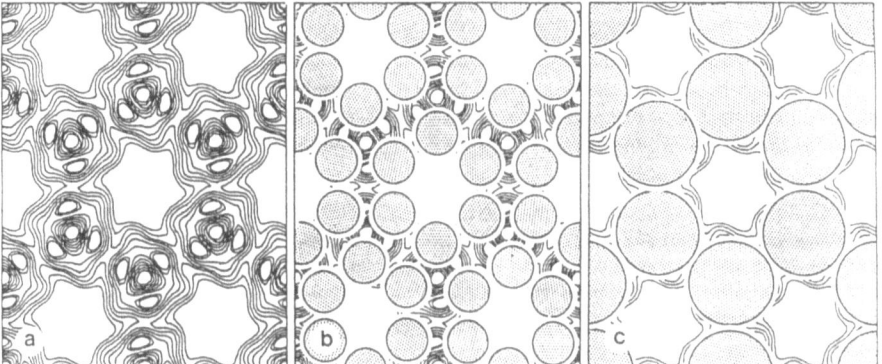

Fig. 10.6. Schematic representation of the mass distribution of a hexagonal S-layer lattice **a** showing the regular arrangement of immobilised small **b** and large **c** molecules. The macromolecules are bound at geometrically well defined locations since functional groups on the lattice are located on equivalent positions in defined orientations. Depending on the size and morphology of the immobilised molecules a dense packing is possible while pores at the same time remain open.

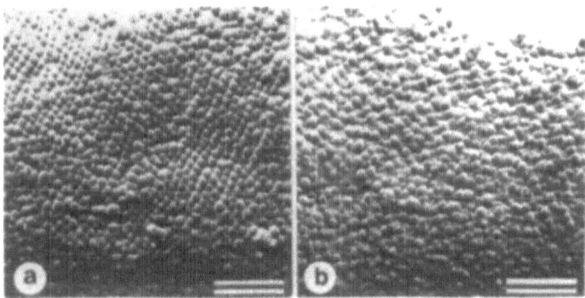

Fig. 10.7. Electron micrographs of freeze-etched preparations of S-layers of *Clostridium thermohydrosulfuricum* L111-69 after covalent binding of ferritin molecules. **a** After activating the carboxyl groups on the S-layer protein with carbodiimide, ferritin is covalently bound in maximum dense packing order reflecting the underlying hexagonal array. **b** Ferritin is bound to the carbohydrate chains of the S-layer glycoprotein after activation with cyanogen bromide. The immobilised ferritin molecules are densely packed but exhibit less ordered arrangement compared to a). Bars, 100 nm.

Immobilisation of Macromolecules through Covalent Bonds

S-layers have been shown to be particularly suitable as a matrix for the controlled covalent attachment of foreign molecules. Generally the carboxyl groups on the S-layer protein or the hydroxyl groups of the carbohydrate chains were activated. The carboxyl groups were most commonly activated with carbodiimide whereas the hydroxyl groups were treated with cyanogen bromide (Sára *et al.* 1988, Sára and Sleytr 1989). A broad spectrum of macromolecules, including ferritin (Fig. 10.7) or biologically active proteins and ligands (e.g. enzymes, protein A, avidin or biotin) was used (for more detailed information see Sára *et al.* chapter 6 this volume). The spatial distribution of immobilised molecules smaller than the constituent S-layer subunits revealed that more than one carboxyl group per S-layer subunit must be available for binding.

Covalently linked molecules frequently exhibit a less ordered arrangement on S-layer lattices than molecules immobilised by ionic bonds. This difference can be explained by the fact that most activation reactions lead to an increase in the number of potential binding sites.

S-layers as Immobilisation Matrices in the Development of Biosensors

Current Biosensor Technologies

Research activities involved in the development of biosensors are multidisciplinary, encompasing such diverse areas as biochemistry, physics and material sciences (Turner *et al.* 1988, Schmid and Scheller 1989, Cass 1990, Hall 1990). Biosensors are a comparatively new generation of analytical devices in which the key element is a sensing layer of biologically active material. This layer is also called transmitter since it confers on the sensor a high specificity to a target analyte in a complex medium. There is a wide range of biological components that can be used to form a biosensor. Enzymes which act as biological catalysts are by far the most commonly used. The sensing layer is combined with a transducer which converts the biological reaction into a physically exploitable signal. Most of the presently known transduction principles require an intimate contact between the sensing layer and the transducer.

The electrochemical transduction of enzyme reactions by amperometric detection involves the oxidation or reduction of the enzymatic reaction products at a metallic working electrode. The working electrode is held at a constant potential with respect to a reference electrode. The potential is chosen so that the species of interest is either oxidised or reduced at the working electrode. This causes a transfer of electrons resulting in a current which is directly proportional to the concentration of the analyte at the electrode surface. Although the transduction principle is simple the sensing layer must show a high activity and be in close contact with the electrode surface in order to obtain high signals. This is because only those enzymatic reaction products undergo the electrochemical transduction which reach the electrode surface by diffusion. By far the most prominent application of amperometry is the monitoring of glucose. Glucose oxidase converts glucose to gluconic acid and hydrogen peroxide (H_2O_2). The amount of hydrogen peroxide is proportional to the glucose concentration and can be detected by electrochemical oxidation. Problems with this system include the dependence of the response to the initial oxygen concentration and the high potentials (+550 mV against Ag/AgCl-reference electrode) required to monitor hydrogen peroxide. Thus the possibility of interference from other electroactive species in solution is high. An elegant method of avoiding a lot of the problems associated with the detection of hydrogen peroxide is the direct transfer of electrons from the enzyme to the sensing electrode. Unfortunately, satisfactory direct electron transfer has proved extremely difficult to achieve for all but a few enzymes. Due to the large distances the tunneling of the electrons from the active site of the enzyme to the electrode surface is highly unfavourable. One solution is the use of electron mediator molecules which shuttle the electrons from the redox center of the enzyme to the electrode. (Cass *et al.* 1984, Degani and Heller 1987, Degani and Heller 1988, Foulds and Lowe 1988).

Especially optically based transduction principles (for a detailed description see Wolfbeis 1988) require thin and biologically active sensing layers. When light undergoes a total internal reflection at an optical interface, a decay of energy away from the point of reflection into the surrounding medium occurs (Born and Wolf 1980). This energy field is known as an evanescent wave and extends into the medium around the optical interface for a distance similar to the wavelength of light. Due to the exponential decay of the energy field, only optically active compounds in the intimate vicinity of the interface will be detected contrary to compounds in distant regions which are not illuminated. As an example, the evanescent wave technique can be used to excite and measure the fluorescence of molecules bound to the surface of an optical fibre or waveguide. This technique is called "Total Internal Reflection Fluorescence". It is a common technique for optical immunosensors (Schultz 1988). The optical fibre or waveguide is coated with antibodies which specifically trap antigens labeled with fluorescent markers. The restricted energy field of the evanescent wave allows for differentiation between fluorescent markers in the medium and those bound to the sensing layer since only the latter ones are illuminated. In order to obtain good signal to noise ratios the amount of bound antibodies must be high. Contrary to evanescent wave techniques, enzymes and fluorescent dyes which are attached at the front end of an optical fibre are illuminated directly. This technique may be used when fluorescent markers are bound to the sensing layer *a priori* and unspecific fluorescence from the medium is negligible (Schultz 1988). Although higher signals are obtainable compared to evanescent wave techniques, the thickness of the sensing layer must be small and the activity high in order to obtain high signal-to-noise ratios and short response times.

Although this brief overview only referred to some of the most common transduction principles it becomes obvious that there is a need to develop thin immobilisation supports with high binding capacities for obtaining biosensors with higher sensitivity, shorter response times and wider linear ranges.

Conventional and Novel Immobilisation Supports

Electrodes for electrochemical biosensors are commonly made of inert metals, such as platinum or gold, or carbon. The latter is either available in the form of graphite, glassy carbon or pyrolytic graphite, as a solid or a paste. The fabrication of an enzyme sensor using carbon paste as electrode material is done by mixing the enzyme with graphite powder. The electrochemical reaction only takes place on the solid-liquid interface where the analyte reaches the sensing layer. Usually it is necessary to cover the carbon surface with a membrane because the enzymes are only physically trapped and not covalently bound. They would otherwise diffuse into the medium. In this way, the total activity may be increased by encaging more enzyme in the space between the electrode surface and the membrane (Bartlett 1987). Enzymes may be covalently bound to inert materials after introducing reactive groups on the surface. As an example, silanisation by tip coating is a commonly used technique. Although this technique does not have the problem with solubilisation of the enzyme into the medium, the activity of the sensing layer is usually low (10^{-12} mol enzyme/cm^2). Nevertheless, platinum, gold or carbon are amenable to many advanced manufacturing techniques, allowing inexpensive mass production. These techniques comprise thin- and thick film technologies which are known from the electronic industry. Optical fibres or waveguides are also prepared for a covalent attachment of molecules by silanisation. The density of reactive groups is comparable to that on metals.

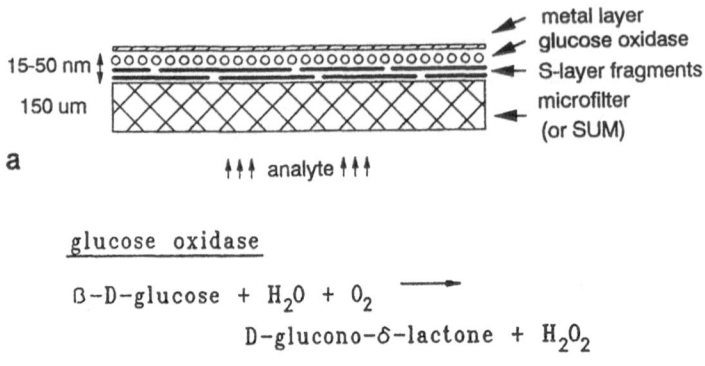

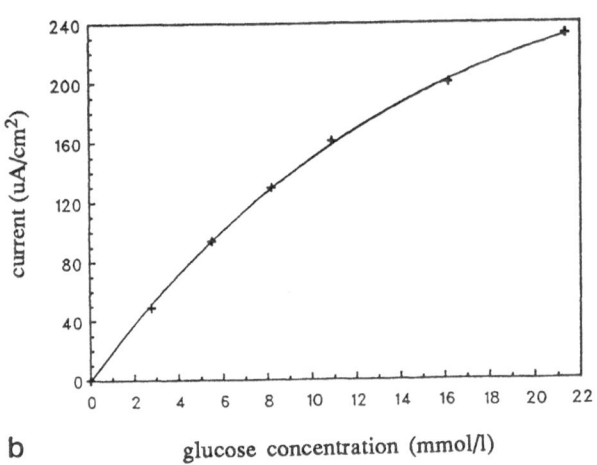

Fig. 10.8. a Schematic illustration of a single enzyme S-layer biosensor. In the fabrication of a glucose sensor, glucose oxidase molecules are covalently bound to the exposed S-layer surface of a S-layer ultrafiltration membrane (SUM). The membrane consists of S-layer fragments which had been deposited on a microfilter and crosslinked with glutaraldehyde. The electrical contact to the enzyme molecules is established by depositing a platinum or gold layer on the 15-20 nm thick sensing layer. The analyte reaches the sensing layer through the open structure of the microfilter. **b** Reaction scheme and typical response curve of the S-layer glucose sensor illustrated in a). The output signal is plotted against the glucose concentration. The linearity range reaches 20 mmol glucose/l solution.

In comparison to conventional immobilisation techniques, crystalline S-layers show several advantages as immobilisation matrices for the development of biosensors. Their binding capacity is two to ten times higher compared to conventional supports (Sára and Sleytr 1989). In additon they function as isoporous molecular sieves. This characteristic feature can be used for building protective layers which prevent disturbing components from reaching the biologically active molecules immobilised on the S-layers (Sleytr and Sára 1985, Sleytr and Sára 1988).

Single Enzyme (Amperometric) S-layer Biosensor

The first S-layer based biosensor developed was a conventional amperometric glucose sensor using glucose oxidase as the biologically active component. This system was chosen because many comparative data on the stability and activity of this sensor were available from the literature. The working electrode consists of a S-layer ultrafiltration membrane (SUM) which is made of a microfilter coated with crosslinked S-layer fragments and a thin layer of glucose oxidase molecules. The enzyme is bound to the carbodiimide activated carboxyl groups of the S-layer protein (Fig. 10.8a). In order to facilitate the electrochemical oxidation of hydrogen peroxide in close vicinity to the glucose oxidase molecules the enzyme layer is either pressed against a metal plate or metal coated by sputtering with platinum or gold. The high degree of preservation of enzyme activity which is 50%-70% of the activity of the immobilised enzyme, high packing density and short diffusion lengths yield high signal levels (150nA/mm^2/mmol glucose), a wide linearity range (up to 20 mmol glucose) and fast response times (10 s -30 s) (Fig. 10.8b). The sensor can be used for more than 24 h continuously and stored for at least six months. Measurements with this electrode design have shown that the oxygen partial pressure is not influenced by variations in the oxygen tension in the medium. This phenomenon is explained by the microenvironment which is established by the space in between the densely packed enzyme molecules, the porous S-layer matrix and the closely adhering metal coating.

Multi Enzyme (Amperometric) S-layer Biosensor

S-layer ultrafiltration membranes are only suitable as immobilisation supports for single enzyme sensors. Simultaneous immobilisation of several different enzyme species leads - similar to conventional immobilisation matrices (see above) - to an uncontrollable competition for the available binding sites and consequently to an unequal loading. In a cocktail of several enzymes with different sizes the small molecules have been shown to predominate over the bigger ones in occupying the binding sites. Therefore, the development of a multienzyme sensor required a modification of the fabrication process. Enzymes are immobilised on S-layer fragments or S-layer self-assembly products in suspension before the deposition and fixation on a microfilter or alternatively on a S-layer ultrafiltration membrane (SUM). Fig. 10.9a and b illustrate the fabrication process of a multi enzyme S-layer biosensor. The different enzyme labeled S-layers may be either deposited sequentially (a) or as a mixture (b) on the microfilter. This procedure allows an individual optimisation of the immobilisation parameters for the individual enzymes and an accurate adjustment of the amounts of enzyme with respect to their activity. Furthermore, this method leads to a well defined multilayer structure of porous enzyme layers where protective S-layers as charged or uncharged isoporous molecular sieves can be integrated. Although such a multienzyme sensor may consist of many superimposed S-layers the total thickness of the sensing layer remains small in comparison to layers produced by conventional manufacturing techniques.

On the basis of this technology several different multienzyme S-layer biosensors such as a sucrose sensor comprising a three enzyme cascade consisting of immobilised fructosidase, mutarotase and glucose oxidase were developed (Fig. 10.10).

When a glucose electrode is immersed in biological fluids (e.g. blood, urine), other substances like ascorbic acid or uric acid can be oxidised at the working electrode,

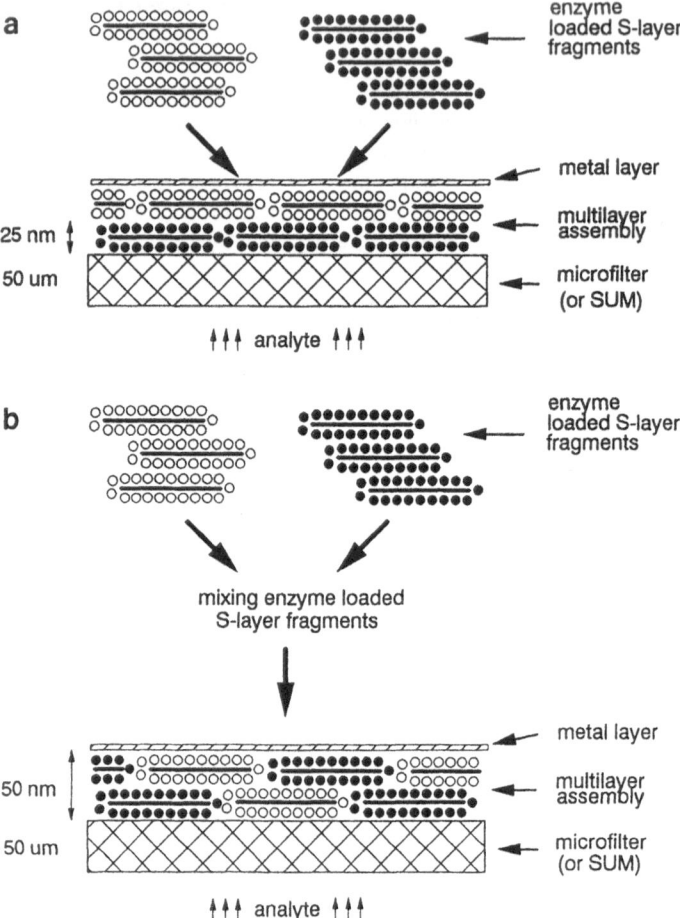

Fig. 10.9. Schematic illustration of the fabrication process and construction principle of a multienzyme S-layer biosensor: The multilayer assembly may be either produced by depositing the individual enzyme loaded S-layers sequentially as shown in a or by mixing them b prior to the deposition on the microporous support. In both cases the multilayer assembly can be crosslinked with glutaraldehyde and is subsequently metal coated with platinum or gold. The contribution of a single enzyme layer to the thickness of the sensing layer is approximately 25 nm. The analyte reaches the multienzyme assembly through the microfilter. Instead of the microfiltration membrane an S-layer ultrafiltration membrane (SUM) can be used as deposition matrix.

yielding an undesirable output and biased results. The cross-sensitivity to ascorbic acid could be eliminated by adding a protective layer of S-layer fragments loaded with ascorbate oxidase at the front end of the sensor (Fig. 10.11). In this assembly ascorbic acid is eliminated and therefore has no further effect on the monitoring of glucose.

The method of depositing enzyme loaded S-layer fragments onto microfiltration membranes (Fig. 10.9) can also be used for the production of single enzyme sensors. This technique has the advantage of binding a higher amount of enzyme per (projected)

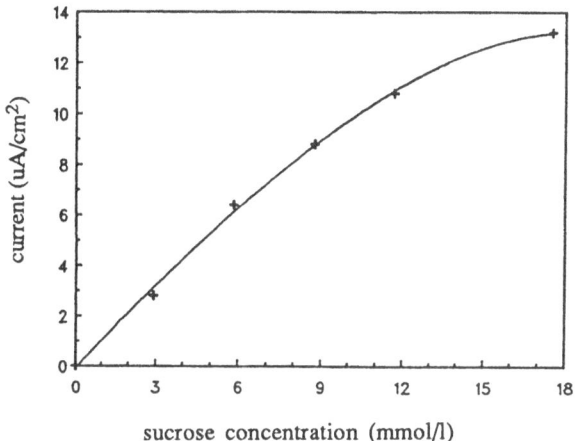

β-fructosidase, mutarotase, glucose oxidase

sucrose + H_2O ⟶ α-D-glucose + β-D-fructose

α-D-glucose ⟷ β-D-glucose ⟶

D-glucono-δ-lactone + H_2O_2

Fig. 10.10. Reaction scheme and characteristic response curve of a three-enzyme (fructosidase-mutarotase-glucose oxidase) S-layer sucrose sensor. The output signal is plotted against the sucrose concentration. The linearity range reaches a value of 12 mmol sucrose/l solution.

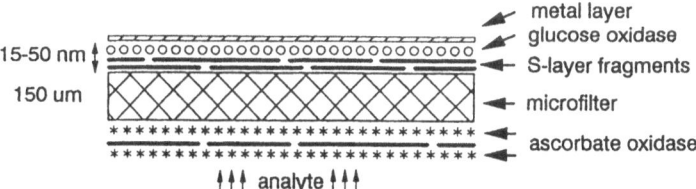

Fig. 10.11. Schematic drawing of a glucose sensor which is not disturbed by the presence of ascorbic acid. The glucose sensing layer is fabricated as shown in Fig. 10.8a. The ascorbate oxidase loaded S-layers are subsequently deposited at the front end of the microfiltration membrane. With this sensor ascorbic acid is eliminated at the front end of the sensor.

unit area at the cost of somewhat thicker sensing layers. Such multilayer single enzyme sensors have been shown to be less sensitive to the metal coating since only the outermost enzyme layer may be destroyed while the inner layers remain unaffected. This is explained by the fact that:

i   the bombardment of the metal atoms only affects the enzyme layer exposed on the surface, and

ii the inner layers do not dry completely in the evacuated evaporation or sputter unit.

Based on this technique the following mono- and multienzyme biosensors were developed:

Sensor type	Enzyme(s)
glucose	glucose oxidase
alcohol	alcohol oxidase
xanthine	xanthine oxidase
sucrose	fructosidase, mutarotase and glucose oxidase
maltose	maltase and glucose oxidase
cholesterol	cholesterol esterase and cholesterol oxidase

Optical Biosenors

S-layers as immobilisation matrices can also be used as interfaces between optical fibers or waveguides and molecules of the sensing layer (e.g. enzymes, fluorescent markers). For this type of biosensor it is not S-layer fragments but S-layer protein monolayers recrystallised on transparent supports that are used. Currently a glucose sensor using glucose oxidase and a pH-sensitive fluorescent dye is being developed. The detection principle is based on changes in pH-values in the vicinity of the enzyme due to the production of gluconic acid in the course of the enzyme reaction. In this optical sensor the sensing layer - which consists of superimposed S-layers with the immobilised enzymes and the fluorescent markers - is attached to the front end of the optical fibre.

Concluding Remarks and Future Outlook

S-layers have been shown to be an appealing alternative to the well established immobilisation supports. They are nature tailored patterning elements in the nanometer range. Their constituent subunits are in the size range of most biologically active molecules. The distribution and orientation of functional groups and the physicochemical properties of the crystalline structure is determined by the periodic repetition of a single unit cell in a crystalline array. Binding of functional molecules to S-layers generally does not alter the molecular sieving properties of the protein network. Based on electrochemical detection principles these sensors yield high signal-to-noise-ratios, short response times and a good storage stability. In addition, the fabrication of multienzyme S-layer biosensors has shown that a controlled assembly of multifunctional devices is possible. Photometric S-layer biosensors including immunosensors are currently under development.

Since the arrangement of immobilised molecules frequently follows the crystalline structure of the S-layer lattice we expect that this unique property will lead to new developments in the nanometer technologies. S-layers which have been recrystallised on solid supports may be used as patterning elements for structuring microelectronic or microoptical devices (Sleytr et al. 1988b, Pum et al. 1989b, Sleytr et al. 1989, Fisher et al. 1990, Utsugi 1990). A crystalline protein monolayer is seen as a fundamental structural principle for the development of biomolecular electronic devices. It is knowledge of the location and orientation of adjacent unit cells in crystal lattices which simplifies many problems such as molecular addressability or wiring (Haddon and Lamola 1985, Robinson and Seeman 1987, Schmid and Karube 1988).

S-layers have been shown to be ideally suited for investigations by high resolution electron (Sleytr *et al.*1987, Engelhardt 1987) and scanning force microscopy (SFM) since a broad spectrum of image processing procedures can be applied for the evaluation of periodic arrays (Pum *et al.* 1989a, Stemmer *et al.* 1989, Beveridge *et al.* 1990). The application potential of S-layers for the attachment of molecules for examination by SFM has still to be investigated. Preliminary results have shown that S-layers allow the binding of molecules at well defined positions but can also be used as an internal reference for the calibration of the instrument.

Recently it has been demonstrated that S-layers and S-layer ultrafiltration membranes can be used as carriers for immobilising mono- and multilayers of surfactants (Sleytr and Sára 1985, Sleytr and Sára 1986, Sleytr and Sára 1988, Sleytr *et al.* 1988b). The concept of applying S-layers as isoporous crystalline supports for binding lipid membranes may have considerable impact on the development of new techniques for studying functional aspects of artificial and biological membranes and on new design principles for biosensors. Numerous membrane proteins have the potential to function as selective biological elements based on molecular recognition. So far the major stumbling block for exploiting the application potential of biomembranes and reconstituted membranes is their lower stability in practical devices. Preliminary studies have shown that synthetic and biological membranes maintain their structural and functional integrity for much larger periods of time when attached to S-layers instead of nonstructured supports. These results are not astonishing since many archaebacteria have been shown to possess cell envelopes exclusively composed of a S-layer and a closely associated cytoplasic membrane (for reviews see Sleytr and Messner 1983, Sleytr and Messner 1988, Messner and Sleytr 1992). In particular those species of archaebacteria which dwell under extreme environmental conditions (high temperatures, low pH values, high salt concentrations) possess such envelope structures. Thus, the combination of the structural principles of a porous crystalline

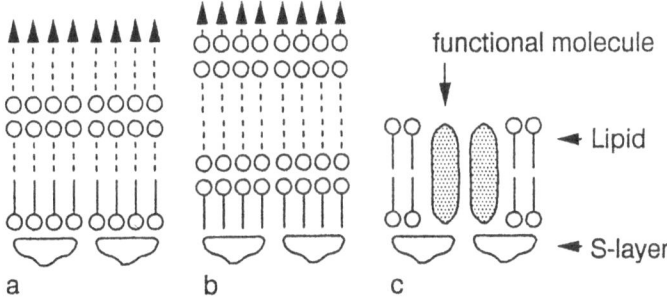

Fig. 10.12. Schematic drawing illustrating the application potential of S-layers as supports for lipid-membranes (e.g. Langmuir Blodgett films). Depending on the hydrophobicity of the S-layer surface monolayers may be deposited either on the S-layer with their hydrophilic head groups **a** or with their hydrophobic tails **b**. In a subsequent step additional monolayers (dotted symbols) can be deposited with their hydrophilic or hydrophobic side. **c** The combination of S-layers and bimolecular lipid layers allows also to incorporate functional molecules.

protein network and a lipid membrane containing functional proteins must have been optimised during billions of years of evolution.

Since the surface properties of S-layers can be adjusted using a broad spectrum of chemical modification reactions more hydrophilic or hydrophobic, lipid films (e.g. Langmuir-Blodgett films) can either bind with their hydrophilic or hydrophobic side (Fig. 10.12a,b). The combination of S-layers with bimolecular lipid membranes where functional molecules (e.g. carriers, ion channels, ionophores, protonpumps, light harvesting and receptor molecules) are incorporated (Fig. 10.12c), opens a broad spectrum of new applications in diagnostics, physiology and biosensor development.

Acknowledegments

The assistance of Angela Neubauer, Claudia Hödl, Andrea Scheberl and Walter Schreier is gratefully acknowledged.

This work was partially supported by grants from the "Österreichischer Fonds zur Förderung der wissenschaftlichen Forschung" Projekt S5705 and S5002 and the Österreichisches Bundesministerium für Wissenschaft und Forschung.

References

Amos LA, Henderson R, Unwin PNT (1982) Three-dimensional structure determination by electron microscopy of two-dimensional crystals. Prog Biophys molec Biol 39:183-231

Andrade JD (1987) Thin organic films of proteins. Thin solid films 152:335-343

Bartlett PN (1987) The use of electrochemical methods in the study of modified electrodes. In: Turner APF, Karube I, Wilson GS (ed) (1988) Biosensors: Fundamentals and applications. Oxford University Press, Oxford, England, pp 211-246

Baumeister W, Engelhardt H (1987) Three-dimensional structure of bacterial surface layers. In: Harris JR, Horne RW (ed) Electron microscopy of proteins, vol 6. Academic Press, London, pp 109-154

Beveridge TJ (1981) Ultrastructure, chemistry, and function of the bacterial wall. Int Rev Cytol 72:229-317

Beveridge TJ, Southam G, Jericho MH, Blackford BL (1990) High-resolution topography of S-layer sheath of the archaebacterium *Methanospirillum hungatei* provided by scanning tunneling microscopy. J Bacteriol 172:6589-6595

Born M, Wolf E (ed) (1980) Principles of Optics. Pergamon Press, Oxford, England, pp 556-592

Carter FL (1983) Molecular level fabrication techniques and molecular electronic devices. J Vac Sci Technol B, vol 1, 4:959-968

Cass AEG (ed) (1990) Biosensors: A practical approach. Oxford University Press, Oxford, England

Cass AEG, Francis G, Hill HAO, Higgins IJ, Aston WJ Plotkin EV, Scott LDL, Turner APF (1984) Ferrocene mediated enzyme electrode for amperometric determination of glucose. Anal Chem 56:667-671

Degani Y, Heller A (1987) Direct electrical communication between chemically modified enzymes and metal electrodes 1. Electron transfer from glucose oxidase to metal electrodes via electron relays, bound covalently to the enzyme. J Phys Chem 91:1285-1289

Degani Y, Heller A (1988) Direct electrical contact between chemically modified enzymes and metal electrodes 2. Methods for bonding electrontransfer relays to glucose oxidase and D-amino acid oxidase. J Am Chem Soc 110:2615-2620

Douglas K, Clark NA (1986) Nanometer molecular lithography. Appl Phys Lett 48:676-678

Eigler DM, Schweizer EK (1990) Positioning single atoms with a scanning tunneling microscope. Nature 344:524-526

Eigler DM, Lutz CP, Rudge WE (1991) An atomic switch realised with the scanning tunneling microscope. Nature 352:600-603

Engelhardt H (1987) Correlation averaging and 3-D reconstruction of 2-D crystalline membranes and macromolecules. In: Mayer F (ed) Methods in microbiology. vol.20, Academic Press, London, pp 357-413

Fisher KA, Yanagimoto KC, Whitfield SL, Thomson RE, Gustafsson MGL, Clarke J (1990) Scanning tunneling microscopy of planar biomembranes. Ultramicroscopy 33:117-126

Foulds NC, Lowe CR (1988) Immobilisation of glucose oxidase in ferrocene-modified pyrrole monomers. Anal Chem 60:2473-2478

Haddon RC, Lamola AA (1985) The molecular electronic device and the biochip computer: Present status. Proc Natl Acad Sci USA 82:1874-1878

Hall EAH (ed) (1990) Biosensors. Open University Press, Ballmoor, England

Hovmöller S, Sjögren A, Wang DN (1988) The structure of crystalline bacterial surface layers. Prog Biophys Molec Biol 51:131-161

Jaenicke R, Welsch R, Sára M, Sleytr UB (1985) Stability and self-assembly of the S-layer protein of the cell wall of Bacillus stearothermophilus. Hoppe-Seyler's Z. Physiol. Chem. 366:663-670

Koval SF (1988) Paracrystalline protein surface arrays on bacteria. Can J Microbiol 34:407-414

Koval SF, Murray RGE (1984) The isolation of surface array proteins from bacteria. Can J Biochem Cell Biol 62:1181-1189

Küpcü Z, März L, Messner P, Sleytr UB (1984) Evidence for the glycoprotein nature of the crystalline cell wall surface layer of Bacillus stearothermophilus strain NRS 2004/3a. FEBS Lett 173:185-190

Messner P, Sleytr UB (1988a) Asparaginyl-rhamnose: A novel type of protein-carbohydrate linkage in a eubacterial surface-layer glycoprotein. FEBS Lett 228:317-320

Messner P, Sleytr UB (1988b) Evidence for the glycoprotein nature of eubacterial S-layers. In: Sleytr UB, Messner P, Pum D, Sára M (ed) Crystalline bacterial cell surface layers. Springer, Berlin, Heidelberg New York, pp 11-16

Messner P, Sleytr UB (1991) Bacterial surface layer glycoproteins. Glycobiology 1:545-551

Messner P, Sleytr UB (1992) Crystalline bacterial cell-surface layers. In: Rose AH, Tempest DW (ed) Advances in microbial physiology, vol 33. Academic Press Inc, London (in press)

Messner P, Pum D, Sára M, Stetter KO, Sleytr UB (1986a) Ultrastructure of the cell envelope of the archaebacteria Thermoproteus tenax and Thermoproteus neutrophilus. J Bacteriol 166:1046-1054

Messner P, Pum D, Sleytr UB (1986b) Characterisation of the ultrastructure and the self-assembly of the surface layer (S-layer) of Bacillus stearothermophilus strain NRS 2004/3a. J Ultrastruct Mol Struct Res 97:73-88

Messner P, Küpcü S, Sára M, Pum D, Sleytr UB (1991) Characterisation and biotechnological application of eubacterial glycoproteins. In: Conradt HS (ed) Protein glycosylation: Cellular biotechnological and analytical aspects. GBF Monographs, vol.15, VCH, Weinheim, FRG, pp 111-116

Pum D, Sára M, Sleytr UB (1989a) Structure, surface charge and self-assembly of the S-layer lattice from Bacillus coagulans E38-66. J Bacteriol 171:5296-5303

Pum D, Sára M, Sleytr UB (1989b) Use of two-dimensional protein crystals from bacteria for nonbiological applications. J Vac Sci Technol B, vol 7, 6:1391-1397

Quate CF (1991) Switch to atom control. Nature 352:571

Robinson BH, Seeman NC (1987) The design of a biochip: A self assembling molecular-scale memory device. Prot Eng 1:295-300

Sára M, Sleytr UB (1987a) Charge distribution on the S-layer of Bacillus stearothermophilus NRS 1536/3c and the importance of charged groups for morphogenesis and function. J Bacteriol 169:2804-2809

Sára M, Sleytr UB (1987b) Molecular sieving through S-layers of Bacillus stearothermophilus strains. J Bacteriol 169:4092-4098

Sára M, Sleytr UB (1987c) Production and characteristics of ultrafiltration membranes with uniform pores from two-dimensional arrays of proteins. J Membr Sci 33:27-49

Sára M, Sleytr UB (1988) Membrane biotechnology: Two-dimensional protein crystals for ultrafiltration purposes. In: Rehm HJ (ed) Biotechnology, vol 6b, VCH Weinheim, FRG, pp 615-636

Sára M, Sleytr UB (1989) Use of regularly structured bacterial cell envelope layers as matrix for the immobilisation of macromolecules. Appl Microbiol Biotechnol 30:184-189

Sára M, Wolf G, Küpcü S, Pum D, Sleytr UB (1988) Use of crystalline bacterial cell envelope layers as ultrafiltration membranes and supports for the immobilisation of macromolecules. In: Dechema Biotechnology Conferences, vol 2, VCH, Weinheim, FRG, pp 35-51

Sára M, Küpcü S, Sleytr UB (1989) Localisation of the carbohydrate residue of the S-layer glycoprotein from Clostridium thermohydrosulfuricum L111-69. Arch Microbiol 151:416-420

Schmid RD, Karube I (1988) Biosensors and bioelectronics. In: Rehm HJ (ed) Biotechnology, vol 6b, VCH Weinheim, FRG, pp 358 360

Schmid RD, Scheller F (ed) (1989) Biosensors: Application in medicine, environmental protection and process control. GBF-Monographs, vol 13, VCH, Weinheim, FRG

Schultz JS (1988) Design of fibre-optic biosenors based on bioreceptors. In: Turner APF, Karube I, Wilson GS (ed) Biosensors: Fundamentals and applications. Oxford University Press, Oxford, England, pp 638-654

Sleytr, UB (1978) Regular arrays of macromolecules on bacterial cell walls: structure, chemistry, assembly, and function. Int Rev Cytol 53:1-64

Sleytr UB, Messner P (1983) Crystalline surface layers on bacteria. Annu Rev Microbiol 37:311-339

Sleytr UB, Messner P (1988) Crystalline surface layers in procaryotes. J Bacteriol 170:2891-2897

Sleytr UB, Messner P (1989) Self assembly of crystalline bacterial cell surface layers (S-layers). In: Plattner H (ed) Electron microscopy of subcellular dynamics. CRC Press Inc, Boca Raton, USA, pp 13-31

Sleytr UB, Sára M (1985) United States Patent Nr. 4,752,395. Structures with membranes having continous pores

Sleytr UB, Sára M (1986) Ultrafiltration membranes with uniform pores from crystalline bacterial cell envelope layers. Appl Microbiol Biotechnol 25:83-90

Sleytr UB, Sára M (1988) United States Patent Nr. 4,849,109. Use of structures with membranes having continous pores

Sleytr UB, Messner P, Pum D (1987) Analysis of crystalline bacterial surface layers by freeze-etching, metal shadowing, negative staining, and ultra-thin sectioning In: Mayer F (ed) Methods in microbiology. vol. 20, Academic Press, London pp 29-60

Sleytr UB, Messner P, Pum D, Sára M (ed)(1988a) Crystalline bacterial cell surface layers. Springer, Berlin, Heidelberg, New York, FRG

Sleytr UB, Sára M, Pum D (1988b) Application potentials of two dimensional protein crystals. In: Paschke F, Fallmann W, Löschner H (ed) Microcircuit Engineering '88, Elsevier Science Publishers B.V. Amsterdam, North Holland, pp 13-20

Sleytr UB, Sára M, Pum D (1989) Application potentials of two dimensional protein crystals. Philips Electron Optics Bulletin 126:9-14.

Smit J (1986) Protein surface layers of bacteria. In: Inouye M (ed) Bacterial outer membranes as model systems. John Wiley & Sons, New York, USA, pp 343-376

Stemmer A, Hefti A, Aebi U, Engel A (1989) Scanning tunneling and transmission electron microscopy on identical areas of biological specimens. Ultramicrosc 30:263-280

Turner APF, Karube I, Wilson GS (ed) (1988) Biosensors: Fundamentals and applications. Oxford University Press, Oxford, England

Utsugi Y (1990) Nanometre-scale chemical modification using a scanning tunneling microscope. Nature 347:747749

Wolfbeis OS (1988) Fiber optical fluorosensors in analytical and clinical chemistry. In: Schuhman SG (ed) Molecular luminescence spectroscopy. Methods and applications: Part 2, Wiley & Sons, New York, pp 129-281

Chapter 11

Immobilised Enzymes in Optical Biosensors

O. S. Wolfbeis

Introduction

Optical sensor technology has emerged in the past 15 years as an attractive alternative to existing sensory schemes such as electrochemistry. This is the result of a number of advantages of optical sensors over others which can include one or more of the following:

i   inertness to electromagnetic interference;
ii the redundance of reference cells;
iii negligible cross talk.

When coupled to optical fibre waveguides, additional features such as remote or in-vivo sensing as well as distributed sensing (i.e. multiple sensing along a fibre) make this technology even more attractive.

Considerable attention has been paid to the development of optical (fibre) sensors for chemical species such as oxygen, pH, carbon dioxide, ammonia, potassium and the like (Wolfbeis 1991a). Development is mainly driven by the demand for such sensors for clinical application, on-line monitoring in-vivo, and in biotechnolgy and environmental applications. Some of the biomedical sensors are manufactured in large quantities now. They are based, in principle, on traditional indicator chemistry and display sufficient selectivity and sensitivity.

However, for determination of biochemical or bio-organic molecules such as glucose, creatinine, cholesterol or lactate, the situation becomes complicated because there are no indicators, carriers or receptors available which would respond to these species with a reversible colour change under conditions at which a sensor usually is operated, that is:

i   at near neutral pH;
ii at room temperature;
iii without addition of reagents;
iv in a continuous way.

In addition, the response should be reversible in order to meet the definition of a sensor. Otherwise, dipsticks may be used, but this is cumbersome if quasi on-line control is desired.

With very few exceptions, the use of enzymes is the only solution at present to overcome the problem of selectively and sensitively recognising target analytes out of a plethora of potentially interfering species. Enzymes therefore have found broad application in analytical sciences, particularly in immobilised form (Guilbault 1984). When immobilised, however, enzymes can display properties different from the respective enzyme in bulk solution. In particular, changes can be observed in the pH optimum, binding constants, reaction rates, long-term stability, selectivity and inhibitor effects. Such effects not necessarily are of the adverse type. Some enzymes are more stable in immobilised than in "free" form.

Table 11.1. Selection of immobilised enzymes that have been used in optical biosensors, and respective transducers

Enzyme	Analyte	Transduction via	Reference
Decarboxylases			
glutamate decarboxylase	glutamate	CO_2	Dremel *et al*. 1991
oxalate decarboxylase	oxalate	CO_2	Schaffar and Wolfbeis 1991
Dehydrogenases			
alcohol dehydrogenase	alcohols	$NADH^a$	Walters *et al*. 1988
glucose dehydrogenase	glucose	$NADH^a$	Narayanaswamy and Sevilla 1988
lactate dehydrogenase	lactate, pyruvate	$NADH^a$	Wangsa and Arnold 1988
Hydrolases			
creatinine iminohydrolase	creatinine	$NH4^+$	Wolfbeis and Li 1991
esterase	various esters	pH	Luo and Walt 1989
urease	urea	$NH3$, $NH4^+$	Rhines and Arnold 1989; Wolfbeis and Li 1991
		pH	Goldfinch and Lowe 1984; Luo and Walt 1989 Yerian *et al*. 1986
penicillinase	penicillin	pH	Goldfinch and Lowe 1984; Kulp *et al*. 1987
Oxidases			
alcohol oxidase	alcohols	O_2	Völkl *et al*. 1980; Wolfbeis and Posch 1988
ascorbate oxidase	ascorbate	O_2	Schaffar 1988
bilirubin oxidase	bilirubin	O_2	Schaffar 1988; Trettnak 1989
cholesterol oxidase	cholesterol	O_2	Trettnak and Wolfbeis 1990
glucose oxidases	glucose	pH	Goldfinch and Lowe 1984; Trettnak *et al*. 1989; Kulp *et al*. 1988
		O_2	Völkl *et al*. 1980; Trettnak *et al*. 1988; Kroneis *et al*. 1987; Schaffar and Wolfbeis 1990; Dremel *et al*. 1989a; Moreno-Bondi *et al*. 1990;
glutamate oxidase	glutamate	O_2	Dremel *et al*. 1991
lactate oxygenase	lactate	O_2	Dremel *et al*. 1989b
lactate mono-oxygenase	lactate	O_2, CO_2	Lübbers *et al*. 1981; Trettnak and Wolfbeis 1989a
phenolase	phenols	O_2	Schaffar and Wolfbeis 1991
sulfite oxidase	sulfite	O_2	Schaffar and Wolfbeis 1991
uricase	uric acid	O_2	Schaffar and Wolfbeis 1991
xanthine oxidase	xanthine	O_2	Völkl *et al*. 1980

[a] Through formation of fluorescent NADH but without regeneration or recycling; response depends on NAD^+ supply.

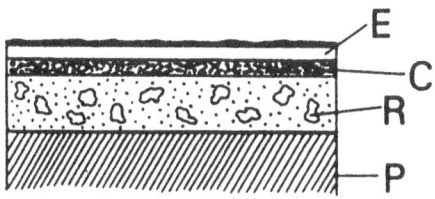

Fig. 11.1. Typical cross section of a sensing layer of an optical biosensor with a chemical transducer. P, polyester mechanical support; R, transducer layer containing a reagent (indicator) sensitive to a species such as pH, oxygen or ammonia which is consumed or produced during an enzymatic reaction; C, optical isolation that prevents ambient light and sample fluorescence to interfere and acts as a mechanical support for the enzyme layer E.

The use of enzymes has tremendously enlarged the sensor field and paved the way for enzyme-based electrochemical and optical biosensors (as opposed to immuno-biosensors which, however, are not sensors for on-line monitoring but rather irreversibly acting single-shot probes). Routine tests based on electrochemical detection are now available for important clinical analytes like glucose, cholesterol, and others (Turner *et al.* 1987; Schmid and Scheller 1989). In such sensors, an electrochemical process is coupled to an enzymatic reaction. Two options exist: In the first (and older) one, a low-molecular species involved in an enzymatic reaction is detected by electrochemical means. Such species include the proton (pH), oxygen or ammonia. In the second option, electrons are directly shuttled from the reaction centre of the enzyme to the (modified) surface of the electrode. This option exists for enzymes of the redox-type in combination with amperometry. It has two advantages: It makes unnecessary the double electron transfer (from the enzyme to water and from hydrogen peroxide to the electrode) and also makes the sensor almost independent of varying oxygen supply.

Given the advantages of optical fibre sensors in certain fields of applications (Schmid and Scheller 1989; Wolfbeis 1991a), it became quite obvious to couple enzymatic reactions to optical chemical transducers. The operating principles of optical chemical sensors have been described in numerous papers, reviews of books and need not be discussed here (Turner *et al.* 1987; Edmonds 1987; Murray *et al.* 1989; Buck *et al.* 1990; Hall 1990; Yacynych and Twork 1990; Blum and Coulet 1991; Wolfbeis 1991a). Basically, a layer composed of immobilised enzyme is placed on top of a chemically sensitive layer such as an oxygen optrode. Fig. 11.1 shows a schematic of the chemically sensitive membranes, while Figure 11.2 shows the optical arrangement for testing their performance.

Four kinds of chemical sensor have mainly been coupled to enzymatic reactions, namely pH, oxygen, CO_2, and ammonia/ammonium. The elegant way of directly shuttling electrons from redox-active enzymes to an electron-accepting sensor (as is the electrode) is, of course, impossible in case of optical (fibre) sensors which are composed of highly insulating materials. However, direct fluorometric detection of NADH formed in all reactions involving dehydrogenases presents an interesting alternative (Table 11.1). No chemical transducer is needed in these cases.

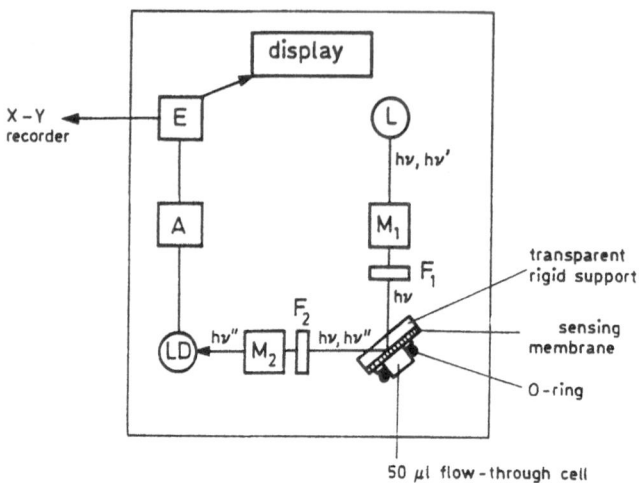

Fig. 11.2. White light (hv, hv') from a source L is monochromatised in M, optionally passes an interference filter F_1, and is shone onto the sensing membrane forming one wall of a small flow-through cell through which sample solutions are pumped. Reflected light (hv) or fluorescent light (hv") passes another filter (F_2) and are registered in light detector LD, then amplified in A, electronically processed in E, and displayed or plotted.

Aspects of Enzyme Immobilisation

The variety of enzymes that conceivably can be combined with optical sensors is large. Table 11.1 lists most of the enzymes that already have been used in optical biosensors thus far. Many of the enzymes given in Table 11.1 have been applied previously in electro-chemical sensors. Because enzymes are unstable outside a cell, the long-term stability of biosensors is often limited and, in average, does not exceed one month of operational lifetime or 200 asays with "real" samples. A rather unusual example of long-term stability is provided by a fibre optic glucose biosensor (Kroneis and Marsoner 1987) based on immobilised glucose oxidase (GOx). After an initial increase in the activity of the sensor due to washout effects of the enzyme gel matrix (resulting in a decreased diffusional barrier for glucose), the sensor activity decreases only slowly over a period of months when properly stored. Similar reports on the stability of immobilised GOx have been published, but the operational lifetime never exceeds three months.

The immobilisation of the enzyme on the transducer surface is known to be a critical step, particularly with respect to sensor reproducibility. In addition, the pH optimum of the enzyme may be broadened and its temperature-activity profile changed. A number of different methods for immobilisation of enzymes on optical sensor surfaces have been described in the literature; some are discussed briefly here with respect to their significance to optical sensing.

Physical Entrapment In this very useful method, the enzyme is incorporated into a hydrogel, such as agar or polyacrylamide, both of which have excellent retaining capability and optical properties. The gel can be fixed to the transducer membrane by using a permeable dialysis membrane (Völkl *et al.* 1980; Kulp *et al.* 1988; Trettnak *et al.* 1989).

Cross-linking The enzyme is cross-linked with a bifunctional species, like glutardialdehyde, to form a gel on either the fibre or a membrane surface (Lübbers *et al.* 1981; Kroneis and Marsoner 1987; Kulp *et al.* 1987; Fuh *et al.* 1988; Rhines and Arnold 1989; Dremel *et al.* 1989a,b; Schaffar and Wolfbeis 1990).

Covalent binding First, the enzyme is fixed covalently to a pre-activated polymer matrix such as a nylon net. The resulting enzyme-loaded membrane is then attached to the transducer to build up a biosensor (Trettnak *et al.* 1988; Trettnak and Wolfbeis 1989ab 1990). Another approach is to bind enzymes covalently to the transducer using classic protein chemistry such as the carbodiimide method (Goldfinch and Lowe 1984) or the avidin-biotin coupling method (Luo and Walt 1989).

Adsorption The enzyme is adsorbed onto supports like kieselgel or activated charcoal which can be attached to the transducer (Wolfbeis and Posch 1988; Schaffar and Wolfbeis 1990). The charcoal acts as both the solid support and an optical isolation that prevents ambient light to interfere.

These procedures have both advantages and disadvantages. After cross-linking the enzyme activity can drop to almost zero. Physical entrapment of enzymes, in contrast, suffers from leakage, and methods for covalent binding are usually complicated and time consuming. Therefore, no general recommendation can be given as to which method is the best. The method of choice depends on the kind of enzyme used, the transducer, and the field of application of the biosensor. Most likely, for each case a specific method for immobilisation must be worked out.

Another critical issue in constructing enzyme-based sensors involves the protection of immobilised enzymes from harmful species in the analyte solution (e.g. microorganisms, free proteases, or inhibitors). Except for inhibitors, a dialysis membrane can largely eliminate such limitations. No solutions have been offered thus far to reduce the influence of inhibitors , except when the enzyme can be shielded from the sample by a polymer (such as silicone rubber) that is permeable to the analyte (e.g. alcohol) but not the inhibitor (Walters *et al.* 1988; Wolfbeis and Posch 1988).

A number of biosensing schemes have been proposed thus far, but this needs comment. Most have been tested with nonreal samples such as stock solutions in defined buffers in well-thermostatted sensor cells. The situation is very much less favorable when measuring real samples because of adverse effects of varying temperature, pH, ionic strength, and inhibitor or promotor concentration. The temperature and pH effects towards the response of a biosensor need to be established independently in each case.

Effects of Enzyme Loading

As with all biocatalytic sensors, enzyme loading and solution conditions are primary factors that must be considered during development of sensors. To a large extent, these factors control the sensor's sensitivity, dynamic range, response time, selectivity, and operational lifetime. Enzyme loading refers to the amount of active enzyme at the active surface. The rate of product generation is governed by the rate of the enzymatic reaction and/or by the rate at which the substrate approaches the biocatalytic layer.

Under high enzyme loading conditions, the rate limiting step is the rate at which the substrate enters the biocatalytic layer. Here, the rate of substrate consumption is mass transport limited and the sensor's response is first order with respect to the substrate concentration. As long as the amount of enzyme activity is sufficiently high, the sensor's response is independent of the actual amount of activity present (Arnold and Wangsa 1991).

Excess enzyme loading of the sensor layer also extends the biosensor lifetime. As the enzyme activity decreases due to natural degradation processes, the sensor still will give the same response until the amount of active enzyme drops below the critical level and substrate mass transport is no longer solely responsible for the kinetics at the sensor tip. For the same reason, sensors with high enzyme loading are less susceptible to the effects of inhibitors and activators.

The kinetics of the enzyme reaction are rate-limiting when enzyme loading is low. Enzyme reaction kinetics in sensors can be described by, e.g. the well-known Michael-Menten equation which relates the initial velocity of the catalyzed reaction to the substrate concentration. If the substrate concentration is much lower than the Michaelis-Menten constant (K_M), the rate of product generation and, consequently, the sensor response, is first order with respect to the substrate concentration. When the substrate concentration is much larger, the sensor response is independent of substrate concentration. Under conditions of low enzyme loading, the K_M value sets the upper limit of detection. A decrease in the dynamic response for a sensor is typically an indication that the activity has decreased significantly and that the rate-limiting process has shifted from being mass transport limited to one which is limited by enzyme reaction kinetics. Overall, high enzyme loading is preferred (Arnold and Wangsa 1991). The Michaelis-Menten kinetics hold for standing samples and require modification when flowing samples are investigated.

Biosensors Based on Oxygen Optrode Transducers

Oxygen sensors, practically all of which are based on quenching of fluorescence by oxygen, can be used as transducers for biosensors with immobilised oxidases which catalyze the following reaction scheme:

$$\text{Substrate} + O_2 + H_2O \xrightarrow{\text{oxidase}} \text{product} + H_2O_2 \qquad (1)$$

Such sensors are usually constructed by placing a layer of immobilised enzyme on top of a chemical optical sensor as schematically shown in Fig. 11.1. In case of such sensors it should be kept in mind that they respond to total oxygen. Hence the response of the sensor to the substrate concentration is strongly affected by the oxygen supply of the analyte solution. Three strategies to overcome problems with varying levels of dissolved oxygen can be applied. One is to measure the oxygen concentration of the analyte solution by a second oxygen sensor (the so-called two-sensor technique; Trettnak et al. 1988); another is based on measurement of the pO_2 gradient in the biosensor with the help of two different indicator species (Lübbers et al. 1981). Thirdly, the oxygen concentration of the analyte solution can be kept constant. This is the case when FIA instrumentation is being used (Schaffar et al. 1989, Dremel et al. 1989a,b).

Glucose Biosensors

Glucose can be determined via the enzymatic oxidation of glucose using glucose oxidase (GOx). Table 11.1 lists the work performed so far. Hydrogen peroxide is usually removed in a second enzymatic reaction employing a catalase. Hydrogen peroxide must be decomposed quickly because of its detrimental effect on the enzyme. Fortunately, the catalase required for this raction is usually present as an impurity in the GOx. When ultrapure GOx fractions are used, the catalase must be added to the enzyme layer to achieve the desired long-term stability of the biosensor. Table 11.2 lists the variety of glucose oxidases exploited so far in optical glucose sensing schemes.

Table 11.2. Comparison of the performance of different optical glucose biosensors based on oxygen transduction

Enzyme source	Immobilisation procedure	Analytical range (mM glucose)	Response time t_{90} (min.)	Stability months	Reference
A. niger	co-crosslinking	0 - 25	0.3	ca 12	Kroneis and Marsoner 1987
A. niger	covalent	0 - 20	5	ca 2	Trettnak et al. 1988
A. niger	adsorption and crosslinking	0 - 2	0.2	ca 6	Schaffar and Wolfbeis 1990
P. amaga-sakiense	adsorption and crosslinking	0 - 4	0.2	ca 4	Schaffar and Wolfbeis 1990
P. nagase	covalent[a]	0 - 2	2	ca 3	Moreno-Bondi et al. 1990
A. niger	covalent	1 - 10	2	ca 3	Moreno-Bondi et al. 1990

[a] on pre-activated membranes; see Table 11.3

Lactate Biosensors

Both lactate mono-oxygenase (LMO) and lactate oxidase (LOx) have been immobilised onto oxygen sensitive layers to obtain lactate sensors (Table 11.1). The use of LMO appears to be advantageous over the use of LOx. This is due to the lack of hydrogen peroxide production during enzymatic action of LMO. A cross section through the sensing layer of an optical lactate biosensor with an oxygen optrode as the transducer is shown in Fig. 11.3.

Table 11.3. Effect of method of enzyme immobilisation onto two kinds of nylon membrane on the performance of two types of lactate sensor (Trettnak and Wolfbeis 1989a)

Lactate Sensor	method 1	method 2
nylon membrane	Immunodyne [TM]	Biodyne [TM]
thickness of indicator layer	10 μm	25 μm
analytical range	2 mM - 50 mM	0.3 mM - 6.0 mM
response times (t_{90})	2.3 min - 3.0 min	2.5 min - 5.0 min
lifetime	1 month	2 months

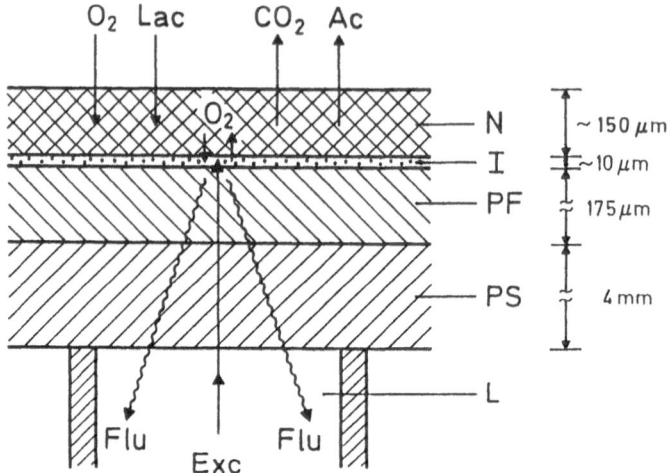

Fig. 11.3. Cross section through an optical lactate sensor membrane. The light guide L carries light from a source to the sensing membrane whose fluorescence of the indicator layer I is monitored. PS is a plexiglass support covered with a polyester foil (PF). They provide mechanical stability because the actual sensing membranes are rather thin. The enzyme is immobilised onto a nylon net (N). Oxygen and lactate diffuse in, CO_2 and acetate out. Oxygen is consumed in N which is registered by the oxygen-sensitive layer I whose fluorescence is quenched by oxygen.

Both the method of immobilisation of LMO and layer thickness strongly affect the response of the sensor (Table 11.3). In addition, the enzyme itself (Table 11.2), as well as the way of its immobilisation, govern the dynamic range of the resulting sensor (Fig. 11.4).

Alcohol Biosensors

The enzymatic determination of alcohols such as ethanol or methanol via oxygen optrodes is usually based on the oxidation of alcohols using an alcohol oxidase (AOx):

$$R - CH_2 - OH + O_2 + H_2O \xrightarrow{\quad AOx \quad} R - CHO + H_2O_2 \qquad (2)$$

The hydrogen peroxide produced is decomposed by catalase, thus preventing the oxidation of the enzyme. Again, the consumption of oxygen is measured and related to the alcohol concentration. The immobilisation of AOx has been performed by incorporation of the enyme into an agar gel, resulting in a sensor with a dynamic range of 0 mM - 10 mM ethanol (Völkl *et al.* 1980). Another design (Walters *et al.* 1988; Wolfbeis and Posch 1988) exploits the capability of alcohols to permeate hydrophobic membranes. An oxygen-sensitive indicator and AOx are immobilised by adsorption onto kieselgel beads suspended in a silicone membrane. The enzyme is protected from inhibitors, micro-organisms, or proteases present in the analyte solution. Due to the poor stability of AOx, the lifetime of this biosensor is rather short, the response being

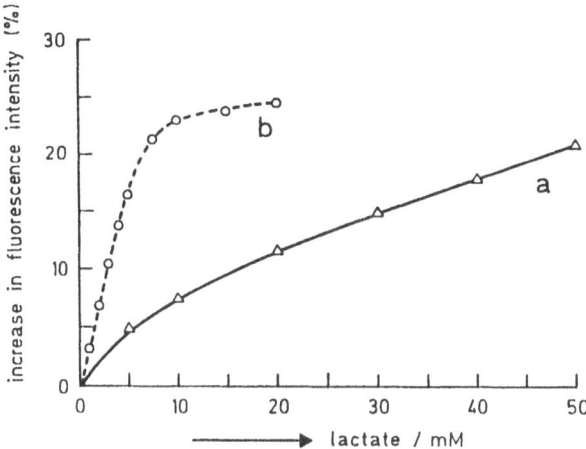

Fig. 11.4. Calibration graphs for the two kinds of lactate sensors specified in Table 11.3. Curve (a), method 1; curve (b), method 2.

reduced to 10 % of its initial activity after two weeks at 4°C. The dynamic range of the sensor is from 10 mM to 500 mM ethanol. All optical sensors using AOx as the enzyme also lack selectivity, because AOx not only catalyzes the oxidation of ethanol, but also of methanol, and even higher alcohols.

Other Types of Enzyme-Based Biosensor Using Oxygen Transduction

Aside from the various kinds of oxidases and oxygenases described so far, several other less common oxidases and oxygenases have been applied in such sensors (Table 11.1). Schaffar and Wolfbeis (1991) report on the immobilisation of various enzymes by the same standard procedure (a cross-linking reaction with glutardialdehyde), and coupling of the enzymatic process to an oxygen transducer based on the quenching of the fluorescence intensity of decacyclene dissolved in silicone rubber (Wolfbeis *et al.* 1985). Only GOx, ascorbic acid oxidase, and tyrosinase display a useful long-term stability which is an important quality criterion for a biosensor. While the data are useful for comparison purposes, they do not imply that glutaraldehyde immobilisation is the most appropriate method. More likely, each single biosensor requires an immobilisation procedure of its own in order to achieve optimal long-term stability of the resulting device.

Biosensors Based on pH Optrodes

pH optrodes can be used as transducers whenever the enzymic reactions causes a change in the pH of the sample. This change can be observed via an immobilised indicator using absorbance, reflection, or fluorescence techniques. A variety of biocatalyzed reactions causing a pH change are known from the literature (Goldfinch and Lowe 1984; Kulp *et al.* 1988; Fuh *et al.* 1988; Trettnak *et al.* 1989; Luo *et al.* 1989) and

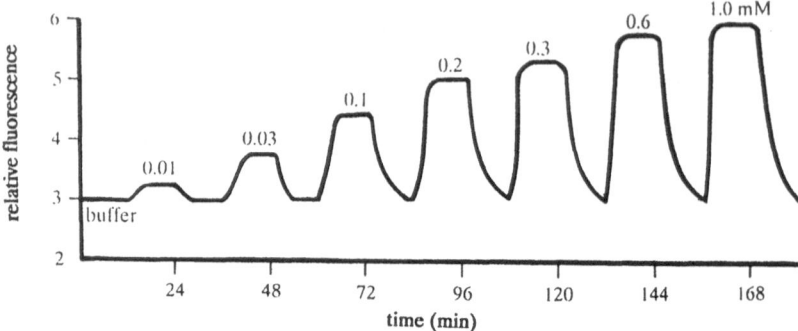

Fig. 11.5. Response of a urea biosensor membrane (based on an ammonium sensor on top of which is immobilised urease) to varying levels of urea in the fluid pumped over the membrane (Wolfbeis and Li 1991).

have been used for biosensing purposes. Typically enzyme reactions include the following:

$$\text{glucose} + O_2 \xrightarrow{\quad\text{GOx}\quad} \text{gluconate} + H^+ + H_2O_2 \qquad (3)$$

$$\text{penicillin} + H_2O \xrightarrow{\quad\text{penicillinase}\quad} \text{penicilloate} + H^+ \qquad (4)$$

$$\text{urea} + 3\,H_2O \xrightarrow{\quad\text{urease}\quad} 2\,NH_4^+ + HCO_3^- + OH^- \qquad (5)$$

A general and severe drawback of using pH optrodes as transducers is the influence of both sample pH and buffer capacity on the measured signal. Another problem is the effect of varying oxygen partial pressure of the sample solution on the response of oxidase type enzymes. Under usual experimental conditions the oxygen partial pressure in the sample can be controlled more easily than buffer capacity. Therefore, optrodes with oxygen transducers usually are preferred. Table 11.1 summarises the kind of sensors in which immobilised enzymes have been coupled to optical pH transduction.

Biosensors Based on Ammonia or Ammonium Optrode Transducers

The enzymatic reaction in Equation 5 has been used to construct optical urea biosensors (Rhines and Arnold 1989). A fluorescein derivative whose fluorescence is pH dependent was dissolved in a sodium and ammonium chloride solution and was held in place at the end of an optical fibre system by means of a microporous teflon membrane. Urease was co-immobilised with BSA via a cross-linking reaction with glutardialdehyde onto the membrane. Ammonia gas formed from urea permeates the Teflon membrane and

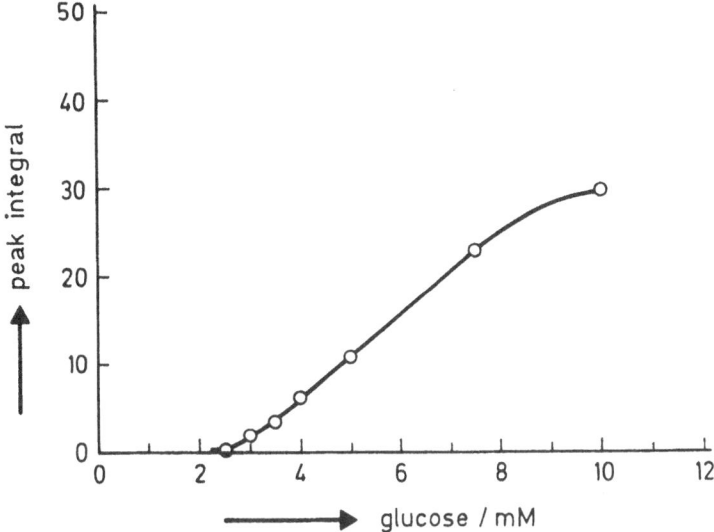

Fig. 11.6. Calibration plot of the optical glucose sensor based on the intrinsic fluorescence of glucose oxidase, as determined in a flow-injection type analyzer so to extend the dynamic range (Trettnak and Wolfbeis 1989c).

changes the pH of the ammonium chloride buffer. This results in a change in the fluorescence intensity of the indicator which, however, shows poor reversibility. The dynamic range of this sensor type was 0.01 mM - 2.5 mM urea, with response times of about 3.3 min.

At near neutral pH, the fraction of free ammonia in the ammonia/ammonium ion equilibrium is rather small, so the detection of ammonium ion (NH_4^+) rather than of free ammonia ($NH3$) was expected to result in a reversible sensor. Indeed, a fully reversible urea sensor is obtained (Wolfbeis and Li 1991) when urease, immobilised on a nylon net, is placed on top of an ammonium-sensitive membrane based on a neutral ion carrier for ammonium ion (Seiler *et al.* 1989). Fig. 11.5 shows the response. Similar results were obtained when immobilised creatinine iminohydrolase was applied in a biosensor for creatinine which is a highly significant parameter in clinical analysis. The limitation of such sensors is in their poor specificity for ammonium over potassium ion.

Biosensors Based on the Intrinsic Optical Properties of Immobilised Enzymes

Enzymes containing flavine type co-enzymes such as FAD reversibly change their optical properties upon interaction with substrates [Trettnak and Wolfbeis 1989b]. Oxidases and oxygenases contain FAD as the co-enzyme which is a fluorescent chromophor. Under 440-nm excitation, FAD and FMN display a green fluorescence of varying intensity (Wolfbeis 1985). Fluorescence is different for the oxidised and reduced form, respectively. This can be exploited for sensing purposes. When, for instance, lactate mono-oxygenase in hydrogel is physically immobilised at the tip of an optical fibre by means of a cellulose membrane, its fluorescence changes as a function of the glucose concentration in contact with the fibre tip. Fig. 11.6 shows the response curve as obtained in a flow-injection type analyzer. Such sensors are simple in construction (though not fabrication) and have the advantage of not requiring a chemical transducer because the enzyme acts as both the recognition and transduction element. However, they have comparatively small dynamic ranges and still depend on oxygen supply.

As can be seen from Table 11.1, some dehydrogenases have been used for sensing purposes as well. They require NAD^+ as a co-substrate which, unlike the flavins FMN and FAD, are not permanently bound to the apo-enzyme. As a result they are difficult to immobilise, although it has been shown that NAD^+ can be bound to poly(ethylene glycol) with its activity being retained (Lammert et al. 1989). Also, NADH (unlike FMN and FAD) does not accept oxygen as a second substrate, so dehydrogenase reactions are driven by the kinetics of a single reaction only (unless coupled to a NADH oxidase system).

Trends

Current problems of (optical) biosensors using immobilised enzymes include thermal lability, loss of activity with time, limited specificity, and high costs. Thermal lability makes difficult shipment of biosensor instrumentation and may be overcome by making use of thermophiles for which there is an obvious need. Alternatively, enzymes may be replaced by more stable bioorganic synthetic receptors (Wolfbeis 1991b). Loss of activity with time results in complicated calibration protocols and, eventually, in complete loss of function. Poor specificity can confine the applicability of enzyme-based biosensors to rather specific situations. Finally, high costs for enzyme preparations may prevent enzyme-based sensors to become mass products.

Acknowledgement

This chapter was prepared on a long and sunny Labor Day weekend in early September 1991 at Georgian Bay (Lake Huron, Canada). I very much acknowledge the hospitality of Karl Raymond Wolfbeis, and his long and stimulating discussions.

References

Arnold MA , Wangsa J (1991) Transducer-Based and Intrinsic Biosensors, in Fiber Optic Chemical Sensors and Biosensors, Wolfbeis OS (ed) CRC Press, Boca Raton, Florida, vol 2

Blum LJ, Coulet PR (1991) Biosensors Principles and Applications. M. Dekker, New York

Buck RP, Hatfield WE, Umana M, Bowden EF (1990) Biosensor Technology, Marcel Dekker, New York - Basel

Dremel BAA, Schaffar BPH, Schmid RD (1989a) Determination of glucose in wine and fruit juice based on a fibre-optic glucose biosensor and flow injection analysis. Anal Chim Acta 225:293 - 299

Dremel BAA, Trott-Kriegeskorte G, Schaffar BPH, Schmid RD (1989b) L-lactic acid determination in milk products based on a fibre optic biosensor and flow injection analysis (FIA). In Biosensors, Application in Medicine, Environmental Protection and Process Control, GBF Monographs Vol 13, Schmid RD, Scheller F (eds) VCH Verlag, Weinheim, pp 225 - 228

Dremel BAA, Schmid RD, Wolfbeis OS (1991) Comparison of two fibre optic L-glutamate biosensors based on the detection of oxygen or carbon dioxide, and their application in combination with flow-injection analysis to the determination of glutamate. Anal Chim Acta 248:351 - 359

Edmonds TE (1987) Chemical Sensors, Blackie, Glasgow - London

Fuh MRS, Burgess LW, Christian GD (1988) Single fiber-optic fluorescence enzyme-based sensor. Anal Chem 60:433

Goldfinch MJ, Lowe CR (1984) Solid-phase opto-electronic sensors for biomedical analysis. Anal Biochem 138:430 -438

Guilbault GG (1984) Analytical Uses of Immobilised Enzymes, Marcel Dekker, New York - Basel

Hall EAH (1990) Biosensors, Open University Press, Buckingham, UK

Kroneis HW, Marsoner HJ (1987) Enzyme sensors using fluorescence based oxygen detection. In Biosensors, International Workshop 1987, GBF monographs Vol 10, Schmid RD, Guilbault GG, Karube I, Schmidt HL, Wingard LB (eds) VCH Verlag, Weinheim, pp 303 - 304

Kulp TJ, Camins I, Angel SM, Munkholm C, Walt DR (1987) Polymer immobilised enzyme optrodes for the detection of penicillin. Anal Chem 59:2849 - 2854

Kulp TJ, Camins I, Angel SM (1988) Enzyme based fiber optic sensors. Proc SPIE (Soc Photoinstrum Engs) 906:134 -139

Lammert R, Ogbono I, Baumeister T, Danzer J, Kittsteiner-Eberle R, Schmidt HL (1989) Poly(ethylene glycol) - NAD$^+$ in amperometric dehydrogenase electrodes with coenzyme recycling. in Biosensors: Application in Medicine, Environmental Protection and Process Control, Schmid RD, Scheller F (eds), GBF monographs Vol 13, VCH Verlag, Weinheim

Lübbers DW, Vvlkl KP, Grossmann U, Opitz N (1981) Lactate measurements with an enzyme optode that uses two oxygen fluorescence indicators to measure the pO_2 gradient directly. Progress in Enzyme and Ion Selective Electrodes, Springer Verlag, Berlin, pp 67 - 73

Luo S,Walt DR (1989) Avidin-biotin coupling as a general method for preparing enzyme-based fiber-optic sensors. Anal Chem 61:1069 - 1073

Moreno-Bondi MC, Wolfbeis OS, Leiner MJP, Schaffar BPH (1990) Oxygen optrode for use in a fiber-optic glucose biosensors. Anal Chem 62:2377 - 2381

Murray RW, Dessy RE, Heineman WR, Janata J, Seitz WR (1989) Chemical Sensors and Micro-instrumentation, ACS Symposium Ser., vol 403, Am. Chem. Soc., Washington, DC

Narayanaswamy R, Sevilla F (1988) An optical fibre probe for the determination of glucose based on immobilised glucose dehydrogenase. Anal Lett 21: 1165 - 1170

Rhines TD, Arnold M A (1989) Fiber-optic biosensor for urea based on sensing of ammonia gas. Anal Chim Acta 227:387 -391

Schaffar BPH (1988) unpublished results

Schaffar BPH, Wolfbeis OS (1990) A fast responding fibre optic glucose biosensor based on an oxygen optrode. Biosensors 5:137 - 142

Schaffar BPH, Wolfbeis OS (1991) Chemically Mediated Fiber Optic Biosensors. In Biosensors Principles and Applications, Blum LJ, Coulet PR (eds), M. Dekker, New York, chapter 8, pp. 163 -194

Schaffar BPH, Dremel BAA, Schmid RD (1989) Ascorbic acid determination in fruit juices based on a fibre optic ascorbic acid biosensor and flow injection analysis. In Biosensors, Application in Medicine, Environmental Protection and Process Control, GBF monographs Vol 13, Schmid RD, and Scheller F (eds) VCH Verlag, Weinheim, pp 229 - 232

Schmid RD, Scheller F (1989) In Biosensors: Application in Medicine, Environmental Protection and Process Control, GBF monographs Vol 13, VCH Verlag, Weinheim.

Seiler K, Morf WE, Rusterholz B, Simon W (1989) Design and characterisation of a novel ammonium ion-selective optical sensor based on neutral ionophors. Anal Sci. 5:557 - 561

Turner APF, Karube I, Wilson GS (1987) Biosensors: Fundamentals and Applications, Oxford University Press, Oxford

Trettnak W (1989) Optical Biosensors Based on Immobilised Enzymes, PhD Thesis, KF University, Graz.

Trettnak W, Wolfbeis OS (1989a) A fibre optic lactate biosensor with an oxygen optrode as the transducer. Anal Lett 22:2191 - 2196

Trettnak W, Wolfbeis OS (1989b) A fully reversible fiber optic lactate biosensor based on the intrinsic fluorescence of lactate mono-oxygenase. Fresenius Z Anal Chem 334:427 - 430

Trettnak W, Wolfbeis OS (1989c) A fully reversible fibre optic glucose biosensor based on the intrinsic fluorescence of glucose oxidase. Anal Chim Acta 221:195 - 203

Trettnak W, Wolfbeis OS (1990) A fiberoptic cholesterol biosensor with an oxygen optrode as the transducer. Anal Biochem 184:124 - 130

Trettnak W, Leiner MJP, Wolfbeis OS (1988) Fibre optic glucose biosensor with an oxygen optrode as the transducer. Analyst 113:1519 - 1523

Trettnak W, Leiner MJP, Wolfbeis OS (1989) Fibre-optic glucose sensor with a pH optrode as the transducer Biosensors 4:15 - 21

Vvlkl KP, Opitz N, LIbbers DW (1980) Continuous measurement of concentrations of alcohol using a fluorescence photometric enzymatic method. Fresensius Z Anal Chem 301:162 - 167

Völkl KP, Grossmann U, Opitz N, Lübbers DW (1984) The use of the O_2-optode for measuring substances such as glucose by using oxidative enzymes for biological applications. In Adv Physiol Sci Vol 25: Oxygen Transport to Tissue, Kovach AGB, Dora E, Kessler M, Silver IA (eds), pp 99 - 100

Walters BS, Nielsen TJ , Arnold MA (1988) Fiber optic biosensor for ethanol, based on an internal enzyme concept. Talanta 35:151 - 157

Wangsa J, Arnold MA (1988) Fiber optic biosensors based on the fluorometric detection of NADH. Anal Chem 60:1080 - 1084

Wolfbeis OS (1985) The Fluorescence of Organic Natural Products, in Molecular Luminescence Spectroscopy: Methods and Applications, Schulman SG (ed), Wiley and Sons, New York, vol. 1, chapter 3

Wolfbeis OS (1991a) Fiber Optic Chemical Sensors and Biosensors, CRC Press, Boca Raton, Florida, vols 1 and 2

Wolfbeis OS (1991b) Optical sensing based on analyte recognition by enzymes, carriers and molecular interactions. Anal Chim Acta 250:181 - 201

Wolfbeis OS, Li H (1991) LED-Compatible fluorosensor for ammonium ion and its application to biosensing. Proc SPIE (Soc Photoinstrum Engs) vol 1587, in press

Wolfbeis OS, Posch HP (1988) A fibre optic ethanol biosensor. Fresenius Z Anal Chem , 332:255 - 260

Wolfbeis OS, Posch HE, Kroneis H (1985) Fiber optical fluorosensor for determination of halothane and/or oxygen. Anal Chem 57:2556 - 2561

Twork JV, Yacynych AM (1990), Sensors in Bioprocess Control, Marcel Dekker, New York - Basel.Yerian TD, Christian GD, Ruzicka J (1986) Enzymatic determination of urea in water and serum by optosensing flow injection analysis. Analyst 111:865 - 871

Yerian TD, Christian GD, Ruzicka J (1988) Flow injection analysis as a diagnostic technique for development and testing of chemical sensors. Anal Chim Acta 204:7 - 13

Chapter 12

Immobilisation of Macromolecules for Obtaining Biocompatible Surfaces

J. M. Courtney, J. Yu and S. Sundaram

Introduction

The basis of this Chapter is a consideration of approaches utilised to improve the performance of surfaces in situations involving blood contact. The achievement of biocompatible surfaces is a principal feature of research and development relating to biomaterials, where a biomaterial is regarded as a material of synthetic or natural origin, used in contact with tissue, blood or biological fluid. An examination of the literature dealing with thrombosis and artificial surfaces (Forbes and Courtney 1987) reveals that references to the initiation of blood coagulation by foreign surfaces and the differences between the effect of an artificial surface and that of the normal endothelium reach back to 1819, while the application of biomaterials extends over a period of about 100 years (Williams 1987). Improvements in technology and clinical procedures have greatly intensified the long-standing interest in blood interactions with surfaces by increasing the number of biomaterials utilised and the range of applications.

It is not reasonable to expect artificial surfaces to resemble the endothelium and such surfaces will generally exert a strong influence on blood. Consequently, the use of biomaterials in clinical applications in which there is continuous contact of blood normally requires administration of an antithrombotic agent. Therefore, opportunity exists for improving the compatibility of biomaterials and the desire for this improvement covers both long-established procedures, such as the artificial kidney (Nosé 1988; Ringoir and Vanholder 1990) and procedures yet to achieve clinical acceptance, such as the artificial heart (Didisheim et al. 1989; Kambic and Nosé 1991).

Biomaterials for blood-contacting applications are dominated by the use of polymers and options for improving compatibility include polymer synthesis, polymer formulation and polymer modification (Courtney et al. 1989). A potential advantage of polymer modification is that of enhancing compatibility with minimal alteration to

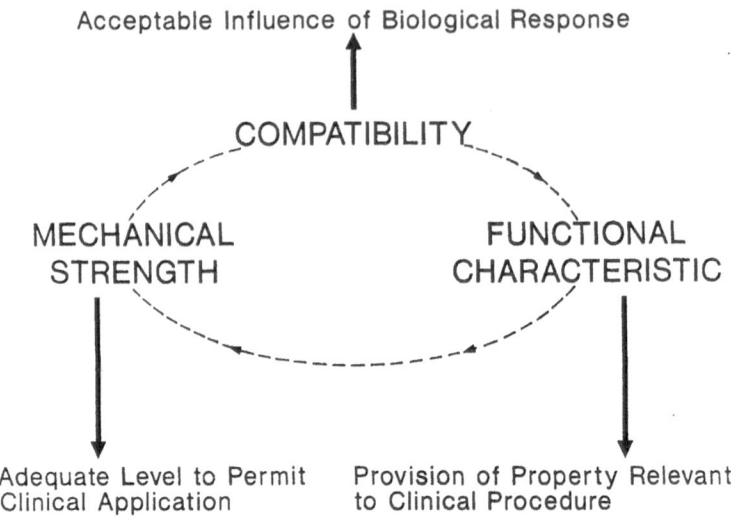

Fig. 12.1. The interacting fundamental properties of a biomaterial

material production and its other properties. Modification by immobilisation of macromolecules offers the possibility of obtaining this desirable property balance. Relevant techniques can be grouped under controlled protein adsorption, the attachment of antithrombotic agents and the preparation of biomembrane-mimetic surfaces. It is appropriate to examine these techniques for modification of the blood response after consideration of biocompatibility and biomaterials and the blood response to biomaterials.

Biocompatibility and Biomaterials

The arrival at an agreed definition of a biomaterial has not been straightforward (Williams 1987) and disagreement remains over what is meant by the term "biocompatibility". While biocompatibility has often been applied to the toxicity of materials as determined by cell culture techniques (Williams 1981), there is a view that biocompatibility should refer only to a clinical evaluation (Klinkmann 1989). However, if biocompatibility is accepted as relating to the behaviour of a material in a biological environment, blood compatibility can be regarded as an aspect of the overall biological response and relevant information can be derived from the knowledge of blood-biomaterial interactions, the influence of materials on blood constituents and the events which induce thrombus formation on artificial surfaces (Andrade et al. 1981; Szycher 1983; Murabayashi and Nosé 1986; Ringoir and Vanholder 1986; Forbes and Courtney 1987; Forbes et al. 1989; Vanholder and Ringoir 1989).

It is important that the blood compatibility of a biomaterial is not considered in isolation from the fact that the biomaterial must possess a functional characteristic in order to perform its intended clinical purpose. This functional characteristic may be flexibility for blood tubing or catheters, solute permeability for a membrane to be used in an artificial kidney, or gas permeability for a membrane to be used in an artificial

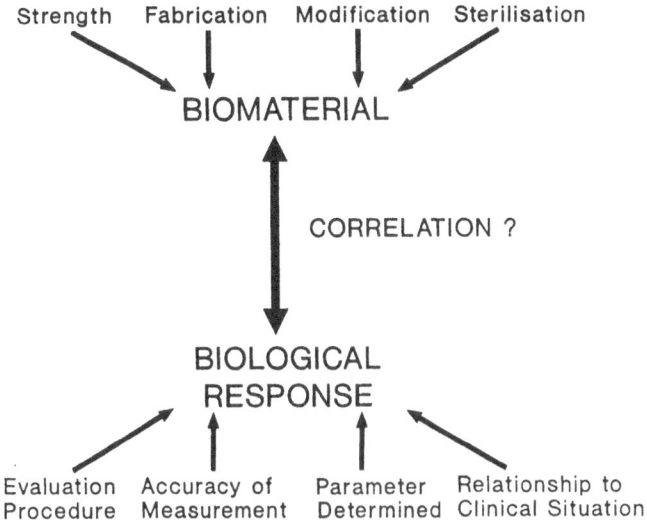

Fig. 12.2. Factors influencing a possible correlation between a biomaterial and a biological response

lung. Since the biomaterial must also possess an acceptable degree of mechanical strength, the fundamental properties of a biomaterial determining its suitability for clinical application can be summarised as compatibility, mechanical strength and functional characteristic (Fig. 12.1). It is essential that techniques for improving the blood compatibility of a biomaterial do not adversely affect the other fundamental properties to an extent that the clinical performance of the biomaterial is unsatisfactory.

Attempts to produce surfaces with enhanced blood compatibility generally utilise a relationship between information on the biomaterial and information on the blood response (Fig.12.2). In practice, a characteristic of the biomaterial is correlated with some feature of the blood response induced by blood-biomaterial contact in order to establish a structure-property relationship.

The usefulness of the correlation is restricted by limitations due to present methods for material characterisation and blood compatibility evaluation. The improved characterisation is important for material development and attention must be given to the refinement of present procedures and the emergence of novel procedures (Ratner 1982, 1983). Problems of blood compatibility evaluation (Courtney *et al.* 1991a) arise from complexity of the interactions of blood with the surfaces of biomaterials.

Blood Response to Biomaterials

It is convenient to consider blood-biomaterial interactions in a manner similar to that normally adopted for an examination of blood coagulation in relation to haemostasis and thrombosis (Mustard and Packham 1977). The processes of haemostasis and thrombosis cover complex reactions between the endothelium, platelets and the coagulation, fibrinolytic and complement systems (Ogston 1983). However, the nature of blood coagulation can be considered in terms of contributions from different aspects, such as platelet activation, intrinsic, extrinsic and common pathways and the control systems involved in thrombus inhibition and fibrinolysis. A similar approach can be used for the events following contact of blood with a biomaterial, although two basic distinctions must be made (Forbes and Prentice 1978). Firstly, the artificial surfaces of biomaterials, in the absence of special modification, cannot undertake the active role in thromboresistance of which the endothelium is capable and, as indicated previously, clinical application may require simultaneous therapy with an antithrombotic agent. Secondly, while the endothelium does not appear to induce protein adsorption under physiological conditions (Bruck 1980), the adsorption of protein is a fundamental step in the interactions of blood with artificial surfaces (Szycher 1983).

In addition to the importance of protein adsorption, knowledge of the relevant aspects of haemostasis and thrombosis indicates that study of the blood response to biomaterials should take into account platelet adhesion, release and aggregation, the coagulation system - in particular the intrinsic pathway - fibrinolytic activity, the complement system and biomaterial influence on erythrocytes and leucocytes.

Protein Adsorption

The adsorption of protein to artificial surfaces occurs rapidly from plasma (Vroman et al. 1972) or whole blood (Gendrau et al. 1981) and the composition of the adsorbed protein layer strongly influences subsequent interactions (Baier et al. 1971; Lyman et al. 1974). Although described as the "conditioning" layer (Baier 1977; Brash 1983; Klinkmann 1984), the adsorbed protein should not be considered "passive" (Bruck 1980), since possibilities exist for transient adsorption, denaturation or changes in conformation (Lee and Hairston 1971; Ihlenfeld and Cooper 1979).

The spontaneous adsorption of protein to artificial surfaces is promoted by the limited solubility of proteins in plasma and by the amphipathic-polar/non-polar-character of protein molecules, which provides a driving force for the concentration of proteins at interfaces (Brash 1983). The nature of the surfaces influences the manner and extent of protein attachment. Electrostatic attachment is important for glass (Chan and Brash 1981), while attachment for polymers may be due to the hydrophobic interaction between non-polar protein and non-polar surface groups in the polar aqueous medium (Brash 1983). Protein adsorption and retention have generally been reported to be greater with hydrophobic than hydrophilic surfaces (Hoffman 1974; Brash et al. 1974; Chuang et al. 1978; Ratner 1981), with conformation changes in adsorbed protein molecules also dependent on hydrophobicity (Absolom et al. 1983).

A focus of blood-biomaterial interactions has been the interest in the relationship between protein adsorption and platelet reactivity, with most evidence acquired for albumin, fibrinogen and gamma globulin. The general conclusion is that platelet adhesion to artificial surfaces is inhibited by adsorption of albumin and promoted by prior adsorption of fibrinogen or gamma globulin (Packham et al. 1969; Salzman et al. 1969; Zucker and Vroman 1969; Lyman et al. 1970; Jenkins et al. 1973; Whicher and Brash 1978; Neumann et al. 1979; Absolom et al. 1979; Adams and Feurstein 1980). The interaction of platelets with adsorbed fibrinogen or gamma globulin has been attributed to the formation of a complex between incomplete heterosaccharides of these proteins and glycosyl transferases located in the platelet membrane (Evans and Mustard

1968; Kim *et al.* 1974; Lee and Kim 1974) by a mechanism similar to that proposed for the platelet-collagen reaction following blood vessel injury (Jamieson 1973). On this basis, the inhibition of platelet adhesion by adsorbed albumin might result from the absence of saccharide chains.

The close relationship between fibrinogen and platelets is demonstrated by the fact that defibrinated or afibrinogenaemic plasma does not support platelet accumulation unless fibrinogen is added (Zucker and Vroman 1969; Mason *et al.* 1971). The presence of fibrinogen is also required for platelet aggregation induced by adenosine diphosphate (Mustard *et al.* 1972). In blood-biomaterial contact, it is possible that the deposition of fibrinogen on artificial surfaces is associated with platelet receptors on adherent platelets and is not independent of platelet deposition (Young *et al.* 1983). Further aspects of the importance of fibrinogen adsorption are the replacement of adsorbed fibrinogen by high molecular weight kininogen (HMWK) (Vroman *et al.* 1980), a protein involved in the activation of the intrinsic coagulation, and the possible interaction of fibrinogen with leucocytes (Szycher 1983).

The adsorption of gamma globulin onto artificial surfaces has been reported to increase platelet adhesion and stimulate the platelet release reaction (Evans and Mustard 1968) and adsorption of this protein may be followed by leucocyte adhesion (Adams *et al.* 1978).

The importance of protein adsorption to blood-biomaterial interactions extends beyond albumin, fibrinogen and gamma globulin and information on the influence of other proteins, including possibly some present in trace amounts, is desirable (Brash 1983).

Protein adsorption is also relevant for the response of blood coagulation factors in that activation of the intrinsic pathways results from interaction of the proteins factor XII, factor XI, HMWK and prekallikrein (Griffin and Cochrane 1979). Activation of the intrinsic coagulation leads to thrombin formation and the rapid production of a fibrin layer on the artificial surface, with the promotion of platelet adhesion and aggregation (Waugh and Baughman 1969; Chuang *et al.* 1979). Thrombin generation also influences the platelet release reaction (Shuman and Levine 1980; Patrono *et al.* 1980; Phillips *et al.* 1980).

Platelet Reactions

The contact of blood with artificial surfaces invariably leads to platelet adhesion and aggregation (Mason 1972; Mason *et al.* 1976; Forbes and Courtney 1987). The adhesion of platelets to protein-coated artificial surfaces produces a change in platelet shape, platelet coalescence into an irregular monolayer and, with increasing platelet adhesion, the formation of mounds with erythrocytes and leucocytes trapped in fibrin (Salzmann *et al.* 1977). Consequences of platelet adhesion are the occurrence of the platelet release reaction (Holmsen *et al.* 1969) in the adhering platelets and platelet aggregation on the artificial surface (Baumgartner *et al.* 1976).

The progress of thrombus formation on an artificial surface is likely to produce interaction between platelets and the intrinsic pathway (Feijen 1977). As observed, thrombin production by the intrinsic pathway induces platelet reaction. Additionally, initiation of the intrinsic coagulation can result from thromboplastins liberated from platelets (Walsh 1982; Needleman and Hook 1982) or from factor XII activation due to platelets stimulated by released adenosine diphosphate.

Erythrocytes

Blood-biomaterial interactions may result in erythrocyte adhesion to the adsorbed protein layer (Feijen 1977) and if haemolysis occurs, released adenosine diphosphate induces the platelet release reaction (Stormorken 1971). Erythrocytes may influence protein adsorption on artificial surfaces (Brash and Uniyal 1976; Uniyal et al. 1982), either by a membrane-related effect or by an effect due to the competitive adsorption of released haemoglobin, and may promote platelet adhesion by reducing the adsorption of platelet-protective proteins or by depositing an adhesive substance (Brash 1983).

Leucocytes

There is leucocyte adhesion to artificial surfaces (Kusserow et al. 1971), with preferential adsorption of polymorphonuclear leucocytes in comparison to lymphocytes (Wright et al. 1978; Lederman et al. 1978; Absolom et al. 1979). The possession by granulocytes of endogenous procoagulant activity (Niemetz 1972; Saba et al. 1973) and proaggregatory activity (Harrison et al. 1966) could promote thrombus formation on artificial surfaces by granulocyte adhesion and by its effect on platelet aggregation (Cumming 1980). The contact of blood with an artificial surface may alter cell function, since leucocyte damage caused by blood-biomaterial contact (Kusserow et al. 1971) produces an impairment in phagocytic activity and a reduced ability to combat infection (Bruck 1980). Leucocyte alterations induced by artificial surfaces may include the release of granulocyte elastase (Courtney et al. 1989) or leucotrienes, interleukins and tissue necrosis factor (Ringoir and Vanholder 1986).

Complement Activation

The influence of artificial surfaces on the complement system has become an important feature of the study of blood-biomaterial interactions (Herzlinger 1983; Chenoweth 1986; Kazatchkine and Carreno 1987). The consequences of complement activation for leucocyte alterations are particularly relevant in the clinical application of biomaterials. Complement activation is believed to mediate leucocyte adhesion to artificial surfaces (Herzlinger and Cumming 1980) and the chemotactic, adhesive and phagocytic responses of polymorphonuclear leucocytes in the inflammatory process, with the anaphylatoxins C3a, C4a and C5a of particular interest.

Fibrinolytic Activity

The integrated nature of the different systems involved in the response of blood to biomaterials makes it necessary to take into account fibrinolysis and fibrinolytic activity. The fact that blood-biomaterial contact induces fibrinolytic activity can be demonstrated by measurement of the levels of fibrin degradation products and activation of the fibrinolytic system has been reported for different clinical applications, including haemodialysis (Kurz et al. 1985) and cardiopulmonary bypass (Bick et al. 1975).

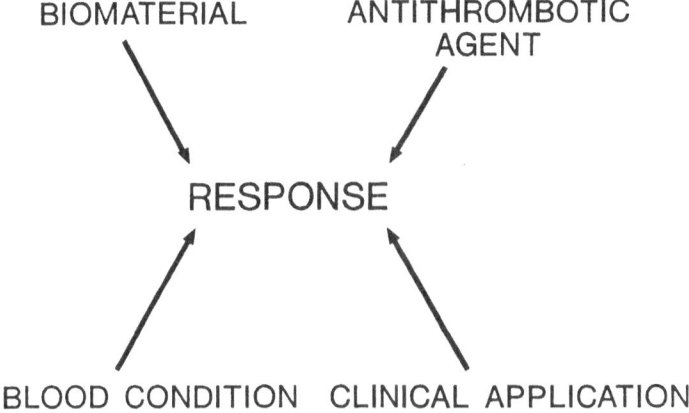

Fig. 12.3. Factors influencing the response of blood to a biomaterial in a typical clinical situation

Modification of the Blood Response

The immobilisation of macromolecules onto a biomaterial surface offers a means of improving compatibility by altering the influence of the biomaterial on the blood response. However, in the clinical situation, the biomaterial is only one of the features determining the response. Other features are the presence of an antithrombotic agent, the condition of the blood and the nature of the clinical application (Fig. 12.3).

The presence of an antithrombotic agent is generally essential to prevent thrombus formation during clinical application and an objective of macromolecule immobilisation is to reduce or eliminate the systemic administration of antithrombotic agents. Blood-biomaterial interactions are influenced by the condition of the blood, which is affected by the disease state and the use of drug therapy. The influence of the clinical application can be considered in broad terms, such as the application time and extent of trauma, or with respect to changes in particular blood components. The *in vitro* behaviour of fibrinogen contrasts with its reactivity in clinical utilisation, where fibrinogen conversion into degradation products and the production of new fibrinogen may occur (Courtney *et al.* 1991b). The platelet response is influenced by diffusion (Feuerstein *et al.* 1975) and shear forces (Richardson *et al.* 1977), with the shear rate and contact time critical factors for platelet adhesion (Rieger 1980). The metabolism of erythrocyte membranes can undergo changes, with the possibility of shear-induced haemolysis (Bruck 1980). Leucocytes resemble platelets in terms of sensitivity to mechanical trauma (Dewitz *et al.* 1977) and shear stress influences leucocyte damage and aggregation (Dewitz *et al.* 1978).

Although the biomaterial is only one of the features contributing to the blood response, alteration to the biomaterial surface is an important option for the improvement of blood compatibility. With respect to the immobilisation of macromolecules, relevant approaches are the treatment of surfaces with protein, the attachment of antithrombotic agents and the preparation of biomembrane-mimetic surfaces.

Treatment of Surfaces with Protein

The alteration to a surface with protein has been strongly influenced by the fact that the adsorption of albumin leads to reduced platelet adhesion. This has been utilised in the identification of polymers capable of albumin adsorption (Lyman *et al.* 1975) or in the preparation of polymers with enhanced albumin adsorption, such as alkyl derivatised polyurethane (Munro *et al.* 1983) and cellulose acetate (Frautschi *et al.* 1983). A reported variation of the technique was the treatment of a polyurethane with an albumin-IgG complex (Mohammad and Olsen 1986).

The efficacy of albumin adsorption is determined by the focus of the blood response and the nature of the application. In the short-term extracorporeal application of haemoperfusion (Giordano 1980), where the passage of blood over sorbents may produce a severe fall in platelets, success has been achieved. The adsorption of albumin to cellulose nitrate-coated activated carbon granules (Chang 1977) or to polystyrene-divinylbenzene resin (Falkenhagen *et al.* 1981) maintained platelet counts at clinically acceptable levels.

In the treatment of surfaces with protein, the term "biolisation" has been introduced (Nosé *et al.* 1971) and is now applied to the chemical and thermal treatment of tissue components, such as proteins, either coated onto a polymer or blended with a polymer (Kambic *et al.* 1983). Therefore, biolised materials include polymers coated with protein, polymers blended with protein, and polymer-protein blends laminated to a base polymer. In contrast to the adsorbed albumin utilised in haemoperfusion, a feature of biolisation is treatment of the modified surface with glutaraldehyde. Biolised materials which have been evaluated include natural rubber treated with albumin or gelatin (Imai *et al.* 1971), polyurethane treated with gelatin (Kambic *et al.* 1978) and a polyolefin elastomer treated with gelatin (Kiraly *et al.* 1977). In applications such as heart assist devices requiring both blood compatibility and the retention of elastomeric properties over extended time periods, biolisation offers an alternative to the synthesis of special polymers by enhancing the blood compatibility of more conventional elastomers.

Attachment of Antithrombotic Agents

Numerous possibilities exist for altering biomaterial surfaces by the attachment of antithrombotic agents. However, efforts have been dominated by the utilisation of heparin and while heparin has been incorporated into polymers (Salyer *et al.* 1971) most procedures have involved attachment of heparin to surfaces by ionic or covalent binding (Gilchrist and Courtney 1980; Kim *et al.* 1983; Fougnot *et al.* 1984; Larm *et al.* 1989; Engbers and Feijen 1991).

Utilisation of Heparin

Heparin (Fig. 12.4) is a negatively charged polysaccharide, mainly composed of alternating residues of sulphated glucoronic, iduronic acid and glucosamine derivatives linked in the 1,4 positions.

The main anticoagulant effect of heparin is that of acting as a catalyst for the inactivation of thrombin, or some other coagulation proteases, by the major inhibitor

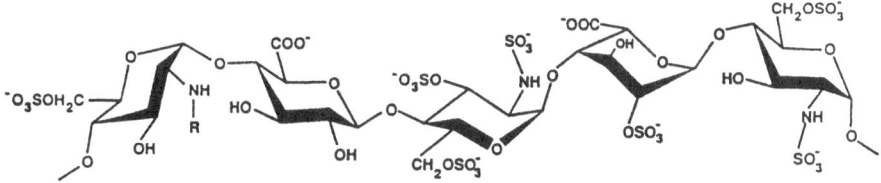

Fig. 12.4. Structure of heparin

antithrombin III (AT-III), which can bind on a specific site on the pentasaccharide sequence of heparin (Lindahl *et al.* 1979; Thunberg *et al.* 1982; Lindahl 1989). In the case of thrombin inhibition, thrombin also binds onto heparin, reacts with bound AT-III, and the inactive complex is then released (Migonney *et al.* 1988). The biological activity of heparin depends on the binding of both thrombin and AT-III and the dominant role of AT-III-binding sites means that these sites should remain unaffected following procedures for surface attachment of heparin (Larm *et al.* 1986).

Because of the strong anionic character of heparin, ionic bonding is readily achieved on cationic surfaces or polymer surfaces rendered cationic by special treatment or copolymerisation. The first procedure for surface heparinisation by ionic attachment (Gott *et al.* 1963) comprised the sequential treatment of a polymer with colloidal graphite, the cationic surfactant benzalkonium chloride and heparin. This procedure is generally unsuitable for flexible materials and these require an elimination of the graphite layer of the coating and an alternative method of providing groups for heparin attachment.

A procedure, avoiding the use of graphite and applicable to a range of polymers, involves contacting the polymer with a suitable cationic surfactant, followed by contact with heparin (Grode *et al.* 1969). Silicone rubber, polyurethane, polyethylene, polypropylene and poly (vinyl chloride) have been treated with a solution of tridodecylmethylammonium chloride (TDMAC) in a petroleum ether- toluene mixture under reflux, followed by dipping in aqueous heparin.

Heparinisation by ionic attachment can be achieved by the incorporation into a polymer of a heparin complex. Evaluation has been made of complexes of heparin with benzalkonium chloride (Fourt *et al.* 1966; Usdin and Fourt 1969), hexadecylpyridinium bromide (Hersh *et al.* 1971), TDMAC (Leininger *et al.* 1972; Grode *et al.* 1972) and cetylpyridinium chloride (Schmer *et al.* 1976). In other procedures for heparinisation by incorporation, an epichlorohydrin-ethylene oxide copolymer was incorporated into poly (vinyl chloride) and reacted with amine and heparin (Falb 1975), and different polymers have been modified by a system consisting of soluble metal salt, a solvent and heparin, with the metal salt polymerising in the treated polymer matrix (Dyck 1972).

The ionic attachment of heparin can be accomplished if polymers contain tertiary amine or quaternary ammonium groups. This approach has been utilised for elastomers (Falb *et al.* 1966; Yen and Rembaum 1971) and copolymers of cellulose acetate (Martin *et al.* 1970), acrylonitrile (Courtney *et al.* 1976), methyl acrylate (Paik Sung *et al.* 1976), methyl methacrylate (Courtney *et al.* 1978) and cellulose (Holland *et al.* 1978; Schmitt *et al.* 1983).

A characteristic of surfaces modified by the ionic attachment of heparin is the removal of the anticoagulant in contact with blood or plasma (Falb *et al.* 1967). This leaching effect has been utilised in the preparation of polymers designed to provide a controlled release of heparin (Tanzawa *et al.* 1973; Mijama *et al.* 1977). Alternatively, heparin removal from surfaces has been reduced by crosslinking the adsorbed heparin with glutaraldehyde (Lagergren and Eriksson 1971; Schmitt *et al.* 1983; Barbucci *et al.* 1985).

Procedures for ionic binding of heparin do not permit long-term retention in contact with blood and the effectiveness of surfaces modified by ionic binding may in fact be dependent on a minimal heparin release rate (Idezuki *et al.* 1975).

It is generally accepted that covalent binding offers the logical approach to long-term retention of heparin and the preparation of surfaces with stability during blood-biomaterial contact. Heparinisation of polymer surfaces by covalent binding normally involves functionalisation, to produce groups such as OH or NH_2, capable of reacting with COOH groups in the heparin molecules.

With classical activation techniques, heparin has been covalently immobilised onto poly (vinyl alcohol) through an acetal bridge (Merrill *et al.* 1970), onto agarose through cyanogen bromide and carbodiimide activation (Schmer *et al.* 1972; Danishefsy and Tzeng 1974), and onto cellulose through radiochemical activation (Hasenfratz and Knaup 1981). The preparation of an amino derivative followed by heparin attachment through cyanuric chloride activation has been applied to silicone rubber (Grode *et al.* 1972), poly (vinyl alcohol) and copolymers of styrene-butadiene and hydroxyethyl methacrylate-glycidyl methacrylate (Peppas and Merrill 1977).

With standard techniques for attachment, surfaces having covalent immobilisation of heparin do not always show an improved blood response (Merrill *et al.* 1970; Hoffman *et al.* 1972). Effectiveness is dependent on the possible utilisation of the functional groups in the active site of heparin molecule in the attachment procedure and inhibition of blood-heparin contact by protein adsorption (Larm *et al.* 1989). In addition, while artificial surfaces with covalently bound heparin are reported to form a complex with AT-III (Fougnot *et al.* 1984), not all such surfaces are able to catalyse the thrombin inhibiting reaction. On the basis that a satisfactory surface must maintain the ability of heparin to bind and activate AT- III (Larsson *et al.* 1980), procedures for covalent attachment for heparin have taken into account the effect of the attachment process on AT-III binding sites.

Protection of the AT-III binding sites during heparin immobilisation has been achieved by a technique producing end-point attachment of heparin (Larm *et al.* 1989). In this technique, heparin is partially degraded with nitrous acid to produce at the reducing terminal residues, fragments with reactive aldehyde groups, which are coupled to an aminated surface by reductive amination. Other functional groups in heparin are not involved in the immobilisation reactions and the AT-III binding sequence is not influenced (Larm *et al.* 1989).

A more controlled heparin immobilisation may require an appropriate spacer between the polymer matrix and the heparin molecule (Ebert and Kim 1982), since the anticoagulant activity of heparin covalently bound to agarose was dependent on the spacer length (Ebert *et al.* 1982a) and hydrophilic poly (ethylene oxide) chains have been preferred as spacers for the covalent attachment of heparin to polyurethane (Park *et al.* 1988), polysiloxane (Grainger and Kim 1988) and polystyrene (Vulic *et al.* 1988) in triblock copolymers.

A variation to heparin immobilisation involves the covalent binding of an albumin-heparin conjugate (Hennink *et al.* 1983), with the objective of obtaining the benefits of both heparin attachment and albumin adsorption.

The successful usage of heparinised materials is dependent on the nature of the application. The focus has been towards catheters (Heyman *et al.* 1985; Eloy *et al.*

1987) and extracorporeal blood purification procedures. In haemodialysis, systems have been evaluated based on ionic binding (Schmer *et al.* 1976, 1977) and covalent binding (Lins *et al.* 1984) but regular application has not yet been achieved. However, interest in such systems for haemodialysis remains because the systemic use of heparin has the potential disadvantages of a risk of haemorrhage (Leonard *et al.* 1969) and an adverse effect on platelets (Lindsay *et al.* 1977; Kelton 1986), while heparinised membranes may not only inhibit platelet reactions but also reduce complement activation. In cardiopulmonary bypass, membrane oxygenators have been heparinised by ionic binding (Rea *et al.* 1972; Hagler *et al.* 1975) but covalent binding (Mottaghy *et al.* 1989; von Segesser and Turina 1989; Nilsson *et al.* 1990; Palatianos *et al.* 1990; Tong *et al.* 1990; Vidern *et al.* 1991) appears to be the approach most likely to gain clinical utilisation.

Utilisation of Platelet Aggregation Inhibitors

The basis for the immobilisation of platelet aggregation inhibitors is the improvement of blood compatibility as a result of reduced platelet adhesion and aggregation rather than inhibition of intrinsic coagulation. Prostaglandin immobilisation (Grode *et al.* 1974) has been demonstrated to produce antiplatelet effects *in vitro* (Ebert *et al.* 1982b), while improved blood compatibility obtained with dipyridamole bound to cellulose has been attributed to the promotion of albumin adsorption or interaction with the enzymatic components of the platelet membrane leading to blocking of the platelet aggregation cascade (Marconi *et al.* 1979). The preferred approach to the immobilisation of platelet aggregation inhibitors is that of covalent binding with options for functionalisation and selection of coupling agents (Ebert *et al.* 1982b; Bamford and Middleton 1983).

Utilisation of Plasminogen Activators

The immobilisation of plasminogen activators is intended to provide surfaces which are fibrinolytically active and capable of reducing thrombus formation by dissolution. Interest has centred on the immobilisation of the enzyme urokinase (Kusserow *et al.* 1973; Sugitachi *et al.* 1980; Ohshiro and Kosaki 1980; Watanabe *et al.* 1981; Aoshima *et al.* 1982; Ohshiro 1983; Senatore *et al.* 1986). Urokinase attachment can be by ionic binding (Aoshima *et al.* 1982) or covalent binding (Watanabe *et al.* 1981).

Biomembrane-Mimetic Surfaces

Procedures for surface modification to improve blood compatibility generally aim at inhibiting some feature of the blood response. An exciting alternative approach proposed by Chapman and coworkers is the preparation of surfaces designed to mimic the biological membrane of blood cells and thereby avoid the recognition by the blood as "foreign".

Investigation of cell membrane structure and function established that erythrocytes and platelets are built upon an asymmetric fluid bilayer of phospholipid (Zwaal *et al.* 1977; Hayward and Chapman 1984) with the phosphorylcholine-PC-head group constituting 88% and 78% of erythrocyte and platelet outer membrane surfaces respectively. PC is present in sphingomyelin and lecithin and contributes to the interfacial and thromboresistant properties of the erythrocyte. The inner membrane surfaces are mainly composed of negatively charged lipid, e.g phosphorylserine (PS), and are thrombogenic. The importance of the PC head group in preventing the activation of blood coagulation was supported by studies demonstrating minimal protein adsorption and platelet activation with a PC- containing lipid, dipalmitoylphosphatidylcholine, representing the outer cell surface, and a thrombogenic response with a negatively charged lipid, dipalmitoylphosphatidylserine, representing the inner cell surface. To overcome the instability of conventional lipids, polymerisable diacetylenic phospholipids containing the PC head group were developed (Durrani and Chapman 1987). These phospholipids have been coated onto substrates by the Langmuir-Blodgett technique (Leaver *et al.* 1983) and stabilised by ultraviolet treatment to provide thromboresistant coatings (Hall *et al.* 1989a).

An approach to achieve stable phospholipid coatings has been to develop PC-containing reactive compounds suitable for covalent attachment to polymer surfaces containing hydroxyl, carboxylic acid or acid chloride groups (Durrani *et al.* 1986; Hayward *et al.* 1986a,b; Hall *et al.* 1989b). This procedure retains the mechanical properties of the substrate, while altering the interfacial properties to mimic those of the cell outer surfaces.

Another technique for the preparation of biomembrane-mimetic surfaces is the preparation of polymers containing the PC head group. Examples are copolymers of 2-methacryloyloxyethyl phosphorylcholine with n-butyl methacrylate (Ishihara *et al.* 1990) or styrene (Kojima *et al.* 1991). Reduced protein adsorption and platelet activation have been reported (Ishihara *et al.* 1991).

Summary

There remains the definite goal of improving the blood compatibility of biomaterials, while retaining acceptable levels of strength and functional characteristics. In this respect, the immobilisation of macromolecules merits further investigation and development, with success influenced by enhanced knowledge of biomaterial characterisation and blood-biomaterial interactions.

Protein adsorption is a key feature both in terms of modifying the blood response and in understanding the mechanism of modification. Meaningful information on alterations to a range of blood constituents is also important for optimisation of modification.

The emphasis on macromolecule immobilisation has been placed on heparinisation and this offers advantages with respect to platelet reactions, complement activation and leucocyte alterations. Progress will be dependent on information relevant to the mode of action. Attention is likely to be directed towards the immobilisation of substances with a direct action on thrombin, such as recombinant hirudin. Other possibilities include immobilisation alone or in combination with heparin of novel platelet aggregation inhibitors or plasminogen activators.

Finally, the development of biomembrane-mimetic surfaces is consistent with the present emphasis on "cell engineering" in biomaterials as targets are set for the preparation of surfaces resembling biological materials in a clinical application.

References

Absolom DR, Neumann AW, Zingg W, van Oss CJ (1979) Thermodynamic studies of cellular adhesion. Trans Am Soc Artif Intern Organs 25: 152-156

Absolom DR, Zingg W, Policova Z, Neumann AW (1983) Determination of the surface tension of protein coated materials by means of the advancing solidification front technique. Trans Am Soc Artif Intern Organs 29: 146-151

Adams AL, Fischer GC, Vroman L (1978) The complexity of blood at simple interfaces. J Colloid Interface Sci 65: 468-478

Adams GA, Feurstein IA (1980) Visual fluorescent and radio-isotopic evaluation of platelet accumulation and embolisation. Trans Am Soc Artif Intern Organs 26: 17-22

Andrade JD, Coleman DL, Didisheim P, Hanson SR, Mason R, Merrill E (1981) Blood materials interactions - 20 years of frustration. Trans Am Soc Artif Intern Organs 27: 659-662

Aoshima R, Kand Y, Takada A, Yamashita A (1982) Sulphonated poly (vinylidene fluoride) as a biomaterial: immobilisation of urokinase and biocompatibility. J Biomed Mater Res 16: 289-299

Baier RE (1977) The organisation of blood components near interfaces. Ann N Y Acad Sci 283: 17-36

Baier RE, Loeb GI, Wallace GT (1971) Role of an artificial boundary in modifying blood proteins. Fed Proc 30: 1523-1538

Bamford CH, Middleton IP (1983) Studies on functionalising and grafting to poly (ether-urethanes). Europ Pol J 19: 1027-1035

Barbucci R, Casini G, Ferruti P, Tempesti F (1985) Surface-grafted heparinisable materials. Polymer 26: 1349-1352

Baumgartner HR, Muggli R, Tschopp TB, Turitto VT (1976) Platelet adhesion, release and aggregation in flowing blood: effects of surface properties and platelet function. Thromb Haemostasis 35: 124-138

Bick R, Schmalhorst W, Crawford L, Holterman M, Arbegast N (1975) The hemorrhagic diathesis created by cardiopulmonary bypass. Am J Clin Pathol 63: 588

Brash JL (1983) Protein adsorption and blood interactions. In: Szycher M (ed) Biocompatible polymers, metals, and composites. Technomic, Lancaster, Pennsylvania, USA, pp 35-52

Brash JL, Uniyal S (1976) Adsorption of albumin and fibrinogen to polyethylene in presence of red cells. Trans Am Soc Artif Intern Organs 22: 253-259

Brash JL, Uniyal S, Samak Q (1974) Exchange of albumin adsorbed on polymer surfaces. Trans Am Soc Artif Intern Organs 20: 69-76

Bruck SD (1980) Properties of biomaterials in the physiological environment. CRC Press, Boca Raton, Florida, USA

Chan BMC, Brash JL (1981) Adsorption of fibrinogen on glass: reversibility aspects. J Colloid Interface Sci 82: 217-225

Chang TMS (1977) Protective effects of microencapsulation (coating) on platelet depletion and particulate embolism in the clinical applications of charcoal haemoperfusion. In: Kenedi RM, Courtney JM, Gaylor JDS, Gilchrist T (eds) Artificial organs. Macmillan, London, England, pp 164-177

Chenoweth DE (1986) Complement activation produced by biomaterials. Trans Am Soc Artif Intern Organs 23: 226-232

Chuang HYK, King WF, Mason RG (1978) Interaction of plasma proteins with artificial surfaces: protein adsorption isotherms. J Lab Clin Med 92: 483-496

Chuang HYK, Crowther PE, Mohammad SF, Mason RG (1979) Interactions of thrombin and antithrombin III with artificial surfaces. Thromb Res 14: 273-282

Courtney JM, Park GB, Fairweather IA, Lindsay RM (1976) Polymer structure and blood compatibility-application of an acrylonitrile copolymer. Biomat, Med Dev Artif Organs 4: 263-275

Courtney JM, Park GB, Prentice CRM, Winchester JF, Forbes CD (1978) Polymer modification and blood compatibility. J Bioeng 2: 241-249

Courtney JM, Robertson LM, Jones C, Irvine L, Douglas JT, Travers M, Ryan CJ, Lowe GDO (1989) Blood compatibility of biomaterials in artificial organs. In: Paul JP, Barbenel JC, Courtney JM, Kenedi RM (eds) Progress in bioengineering. Adam Hilger, Bristol, England, pp 21-27

Courtney JM, Srivatsava S, Robertson LM, Weng D, Lowe GDO (1991a) Biocompatibility assessment: selection of test procedures. In: Paul JP, Rappelsberger P, Schütz PW (eds) The influence of new technologies on medical practice. Verlag für medizinische Wissenschaften Wilhelm Maudrich, Vienna, Austria, pp 163-170

Courtney JM, Irvine L, Jones C, Mosa SM, Sundaram S, McLaughlin KM, Lowe GDO (1991b) Compatibility aspects of biomaterials for artificial organs and assist devices. In: Paul JP, Rappelsberger P, Schütz PW (eds) The influence of new technologies on medical practice. Verlag für medizinische Wissenschaften Wilhelm Maudrich, Vienna, Austria pp 154-162

Cumming RD (1980) Important factors affecting initial blood-material interactions. Trans Am Soc Artif Intern Organs 26: 304-308

Danishefsky I, Tzeng F (1974) Preparation of heparin-linked agarose and its interaction with plasma. Thromb Res 4: 237-246

Dewitz TS, Hung TC, Martin RR, McIntire LV (1977) Mechanical trauma in leukocytes. J Lab Clin Med 90: 728-736

Dewitz TS, Martin RR, Solis RT, Hellums JD, McIntire LV (1978) Microaggregate formation in whole blood exposed to shear stress. Microvascular Res 16: 263-271

Didisheim P, Olsen DB, Farrer DJ, Portner PM, Griffith BD, Pennington DG, Joist JH, Schoen FJ, Gristina AG, Anderson JM (1989) Infections and thromboembolism with implantable cardiovascular devices. Trans Am Soc Artif Intern Organs 35: 54-70

Durrani AA, Chapman D (1987) Modification of polymer surfaces for biomedical applications. In: Feast WJ, Munro HS (eds) Polymer surfaces and interfaces. John Wiley & Sons, New York, USA, pp 189-200

Durrani AA, Hayward JA, Chapman D (1986) Biomembranes as models for polymer surfaces. II. The synthesis of reactive species for covalent coupling of phosphorylcholine to polymer surfaces. Biomaterials 7: 121-125

Dyck MF (1972) Inorganic heparin complexes for the preparation of nonthrombogenic surfaces. J Biomed Mater Res 6: 115-141

Ebert CD, Kim SW (1982) Immobilised heparin: spacer arm effects on biological interactions. Thromb Res 26: 43-57

Ebert CD, Lee ES, Deneris J, Kim SW (1982a) The anticoagulant activity of derivatised and immobilised heparins. Am Chem Soc Adv Chem Ser 199: 161-176

Ebert CD, Lees ES, Kim SW (1982b) The antiplatelet activity of immobilised prostacylin. J Biomed Mater Res 16: 629-638

Eloy R, Belleville J, Paul J, Pusineri C, Baguet J, Rissoan MC, Cathignot P, Ffrench P, Ville D, Tartullier M (1987) Thromboresistance of bulk heparinised catheters in humans. Thromb Res 45: 223-233

Engbers GH, Feijen J (1991) Current techniques to improve the blood compatibility of biomaterial surfaces. Int J Artif Organs 14: 199-215

Evans G, Mustard JF (1968) Platelet-surface reaction and thrombosis. Surgery 64: 273-280

Falb RD (1975) Surface-bonded heparin. In: Kronenthal RL, Oser Z, Martin E (eds) Polymers in medicine and surgery. Plenum Press, New York, USA, pp 77-86

Falb RD, Grode GA, Leininger RI (1966) Elastomers in the human body. Rubber Chem Technol 39: 1288-1292

Falb RD, Takahashi MT, Grode GA, Leininger RI (1967) Studies on the stability and protein adsorption characteristics of heparinised polymer surfaces by radioisotope labelling techniques. J Biomed Mater Res 1: 239-251

Falkenhagen D, Esther G, Courtney JM, Klinkmann H (1981) Optimisation of albumin coating for resins. Artif Organs 5 (Suppl): 195-199

Feijen J (1977) Thrombogenesis caused by blood-foreign surface interaction. In: Kenedi RM, Courtney JM, Gaylor JDS, Gilchrist T (eds) Artificial organs. Macmillan, London, England, pp 235-247

Feuerstein IA, Brophy JM, Brash JL (1975) Platelet transport and adhesion to reconstituted collagen and artificial surfaces. Trans Am Soc Artif Intern Organs 21: 427-434

Forbes CD, Courtney JM (1987) Thrombosis and artificial surfaces. In: Bloom AL, Thomas DP (eds) Haemostasis and thrombosis, 2nd edn. Churchill Livingstone, Edinburgh, Scotland, pp 902- 921

Forbes CD, Prentice CRM (1978) Thrombus formation and artificial surfaces. Brit Med Bull 34: 201-207

Forbes CD, Courtney JM, Saniabadi AR, Morrice LMA (1989) Thrombus formation in artificial organs. In: Paul JP, Barbenel JC, Courtney JM, Kenedi RM (eds) Progress in bioengineering, Adam Hilger, Bristol, England, pp 13-20

Fougnot C, Labarre D, Jozefonwicz J, Jozefowicz M (1984) Modifications to polymer surfaces to improve blood compatibility. In: Hastings GW, Ducheyne P (eds) Macromolecular biomaterials. CRC Press, Boca Raton, Florida, USA, pp 215-238

Fourt L, Schwartz AM, Quasius A, Bowman RL (1966) Heparin-bearing surfaces and liquid surfaces in relation to blood coagulation. Trans Am Soc Artif Intern Organs 12: 155-162

Frautschi JR, Munro MS, Lloyd DR, Eberhart RC (1983) Alkyl derivatised acetate membranes with enhanced albumin affinity. Trans Am Soc Artif Intern Organs 29: 242-244

Gendrau RM, Winters S, Leininger RI, Fink D, Hassler CR, Jakobsen RJ (1981) Fourier transform infrared spectroscopy of protein adsorption from whole blood: ex vivo dog studies. Appl Spectroscopy 35: 353-357

Gilchrist T, Courtney JM (1980) The design of biocompatible polymers. In: Ariëns EJ (ed) Drug design, vol X. Academic Press, New York, USA, pp 251-275

Giordano C (ed) (1980) Sorbents and their clinical applications. Academic Press, New York, USA

Gott VL, Whitten JD, Dutton RC (1963) Heparin bonding on colloidal graphite surfaces. Science 142: 1297-1298

Grainger DW, Kim SW (1988) Poly (dimethylsiloxane)-poly(ethylene oxide)-heparin block copolymers. 1. Synthesis and characterisation. J Biomed Mater Res 22: 231-249

Griffin JH, Cochrane CG (1979) Recent advances in the understanding of contact activation reactions. Semin Thromb Hemostasis 5: 254-273

Grode GA, Anderson SJ, Grotta HM, Falb RD (1969) Nonthrombogenic surfaces via a simple coating process. Trans Am Soc Artif Intern Organs 15: 1-6

Grode GA, Falb RD, Crowley JP (1972) Biocompatible materials for use in the vascular system. J Biomed Mater Res Symp 3: 77-84

Grode GA, Pitman J, Crowley JP, Leininger RI, Falb RD (1974) Surface immobilised prostaglandin as a platelet protective agent. Trans Am Soc Artif Intern Organs 20: 38-41

Hagler HK, Powell WM, Eberle JW, Sugg WL, Platt MR, Watson JT (1975) Five-day partial bypass using a membrane oxygenator without systemic heparinisation. Trans Am Soc Artif Intern Organs 21: 178-185

Hall B, Bird RleR, Chapman D (1989a) Phospholipid polymers & new haemocompatible materials. Angewandte Makromol Chemie 166/167: 169-178

Hall B, Bird RleR, Kojima M, Chapman D (1989b) Biomembranes as models for polymer surfaces. V. Thromboelastographic studies of polymeric lipids and polyesters. Biomaterials 10: 219-224

Harrison MJ, Emmons PR, Mitchell JR (1966) The effect of white cells on platelet aggregation. Thromb Diath Haemorrh 16: 105-121

Hasenfratz H, Knaup G (1981) Improvement of the blood compatibility of cellulosic membranes through the immobilisation of heparin and measurement of biological heparin activity. Artif Organs Suppl 5: 507-511

Hayward JA, Chapman D (1984) Biomembrane surfaces as models for polymer design: the potential for haemocompatibility. Biomaterials 5: 135-142

Hayward JA, Durrani AA, Shelton CJ, Lee DC, Chapman D (1986a) Biomembranes as models for polymer surfaces. III. Characterisation of a phosphorylcholine surface covalently bound to glass. Biomaterials 7: 126-131

Hayward JA, Durrani AA, Lu YC, Clayton CR, Chapman D (1986b) Biomembranes as models for polymer surfaces. IV. ESCA analyses of a phosphorylcholine surface covalently bound to hydroxylated substrate. Biomaterials 7: 252-258

Hennink WE, Feijen J, Ebert CD, Kim SW (1983) Covalently bound conjugates of albumin and heparin. Thromb Res 29: 1-13

Hersh LS, Weetall HH, Brown IW Jr (1971) Heparinised polyester fibres. J Biomed Mater Res Symp 1: 99-104

Herzlinger GA (1983) Activation of complement by polymers in contact with blood. In: Szycher M (ed) Biocompatible polymers, metals, and composites. Technomic, Lancaster, Pennsylvania, USA, pp 89-101

Herzlinger GA, Cumming RD (1980) Role of complement activation in cell adhesion to polymer blood contact surfaces. Trans Am Soc Artif Intern Organs 26: 165-170

Heyman PW, Cho CS, McRea JC, Olsen DB, Kim SW (1985) Heparinised polyurethanes: in vitro and in vivo studies. J Biomed Mater Res 19: 419-436

Hoffman A (1974) Principles governing biomolecule interactions at foreign interfaces. J Biomed Mater Res 8: 77-83

Hoffman AS, Schmer G, Harris C, Kraft WG (1972) Covalent bonding of biomolecules to radiation-grafted hydrogels on inert polymer surfaces. Trans Am Soc Artif Intern Organs 18: 10-17

Holland FF, Gidden HE, Mason RG, Klein E (1978) Thrombogenicity of heparin-bound DEAE cellulose hemodialysis membranes. Am Soc Artif Int Organs J 1: 24-36

Holmsen H, Day HJ, Stormorken J (1969) The blood platelet release reaction. Scand J Haematol Suppl 8: 1-26

Idezuki Y, Watanabe H, Hagiwara M, Kanasugi K, Mori Y, Nagaoka S, Hagio M, Yamamoto K, Tanzawa H (1975) Mechanism of antithrombogenicity of a new heparinised hydrophilic polymer: chronic *in vivo* studies and clinical application. Trans Am Soc Artif Intern Organs 21: 436-448

Ihlenfeld JV, Cooper SL (1979) Transient *in vivo* protein adsorption onto polymeric biomaterials. J Biomed Mater Res 13: 577-591

Imai Y, Tajima K, Nosé Y (1971) Biolised materials for cardiovascular prosthesis. Trans Am Soc Artif Intern Organs 17: 6-9

Ishihara K, Aragaki R, Ueda T, Watenabe A, Nakabayashi N (1990) Reduced thrombogenicity of polymers having phospholipid polar groups. J Biomed Mater Res 24: 1069-1077

Ishihara K, Ziats NP, Tiemey BP, Nakabayashi N, Anderson JM (1991) Protein adsorption from human plasma is reduced on phospholipid polymers. J Biomed Mater Res 25: 1397-1407

Jamieson GA (1973) Role of glycoproteins in platelet function. In: Gerlach E, Moser K, Deutsch E, Williams W (eds) Erythrocytes, thrombocytes, leukocytes: recent advances in membrane and metabolic research. Thieme, Stuttgart, Germany, pp 209-232

Jenkins CSP, Packham MA, Guccione MA, Mustard JF (1973) Modification of platelet adherence to protein-coated surfaces. J Lab Clin Med 81: 280-290

Kambic HE, Nosé Y (1991) Biomaterials for blood pumps. In: Sharma CP, Szycher M (eds) Blood compatible materials and devices. Technomic, Lancaster, Pennsylvania, USA, pp 141-152

Kambic H, Barenburg S, Harasaki H, Gibbons D, Nosé Y (1978) Glutaraldehyde-protein complexes as blood compatible coatings. Trans Am Soc Artif Intern Organs 24: 426-437

Kambic HE, Murabayashi S, Nosé Y (1983) Biolised surfaces as chronic blood compatible interfaces. In: Szycher M (eds) Biocompatible polymers, metals, and composites. Technomic, Lancaster, Pennsylvania, USA, pp 179-198

Kazatchkine MD, Carreno MP (1987) Activation of the complement system at the interface between blood and artificial surfaces. Biomaterials 9: 30-35

Kelton JC (1986) Heparin-induced thrombocytopenia. Haemostasis 16: 173-186

Kim SW, Lee RG, Oster H, Coleman D, Andrade JD, Lentz DJ, Olsen D (1974) Platelet adhesion to polymer surfaces. Trans Am Soc Artif Intern Organs 20: 449-455

Kim SW, Ebert CD, Lin JY, McRea JC (1983) Nonthrombogenic polymers: pharmaceutical approaches. Am Soc Artif Int Organs J 6: 76-87

Kiraly RJ, Arconti R, Hillegass D, Harasaki H, Nosé Y (1977) High flex rubber for blood pump diaphragms. Trans Am Soc Artif Intern Organs 23: 127-132

Klinkmann H (1984) The role of biomaterials in the application of artificial organs. In Paul JP, Gaylor JDS, Courtney JM, Gilchrist T (eds) Biomaterials in artificial organs. Macmillan, London, England, pp 1-8

Klinkmann H (1989) Progress in artificial organs. In: Paul JP, Barbenel JC, Courtney JM, Kenedi RM (eds) Progress in bioengineering, Adam Hilger, Bristol, England, pp 7-12

Kojima M, Ishihara K, Watenabe A, Nakabayashi N (1991) Interaction between phospholipids and biocompatible polymers containing a phosphorylcholine moiety. Biomaterials 12: 121-124

Kurz H, Lemer RG, Weseley S, Nelson JC (1985) Changes in fibrinolytic activity during the course of a single hemodialysis session. Clin Nephrol 24: 1-4

Kusserow B, Larow R, Nichols J (1971) Perfusion-and surface-induced injury in leucocytes. Fed Proc 30: 1516-1520

Kusserow BK, Larow RW, Nichols JE (1973) The surface bonded, covalently crosslinked urokinase surface. Trans Am Soc Artif Intern Organs 19: 8-12

Lagergren HR, Eriksson JC (1971) Plastics with a monolayer of cross-linked heparin: preparation and evaluation. Trans Am Soc Artif Intern Organs 17: 10-12

Larm O, Lins LE, Olsson P (1986) An approach to antithrombosis by surface modification. In: Nosé Y, Kjellstrand C, Ivanovich P (eds) Progress in artificial organs. ISAO Press, Cleveland, pp 313- 318

Larm O, Larsson R, Olsson P (1989) Surface-immobilised heparin. In: Lane DA, Lindahl U (eds) Heparin. Chemical and biological properties, clinical applications. Edward Arnold, London, England, pp 597-608

Larsson R, Olsson P, Lindahl U (1980) Inhibition of thrombin on surfaces coated with immobilised heparin and heparin-like polysaccharides: a crucial non-thrombogenic principle. Thromb Res 19: 43-54

Leaver JA, Alonso A, Durrani AA, Chapman D (1983) The biosynthetic incorporation of diacetylenic fatty acids into the biomembranes of *Acholeplasma laidlawii* A cells and polymerisation of the biomembranes by irradiation with ultraviolet light. Biochim Biophys Acta 727: 327-335

Lederman DM, Cumming RD, Petschek HE, Levine PH, Krinsky NI (1978) The effect of temperature on the interaction of platelets and leukocytes with materials exposed to flowing blood. Trans Am Soc Artif Intern Organs 24: 557-560

Lee WH Jr, Hairston (1971) Structural effects on blood proteins at the gas-blood interface. Fed Proc 30: 1615-1620

Lee RG, Kim SW (1974) The role of carbohydrate in platelet adhesion to foreign surfaces. J Biomed Mater Res 8: 383-388

Leininger RI, Crowley JP, Falb RD, Grode GA (1972) Three years' experience *in vivo* and *in vitro* with surfaces and devices treated by the heparin complex method. Trans Am Soc Artif Intern Organs 18: 312-315

Leonard CD, Weil E, Scribner BH (1969) Subdural haematomas in patients undergoing haemodialysis Lancet 11: 239-240

Lindahl U (1989) Biosynthesis of heparin and related polysaccharides. In: Lane DA, Lindahl U (eds) Heparin. Chemical and biological, clinical applications. Edward Arnold, London, England, pp 159-189

Lindahl U, Bäckström G, Höök M, Thunberg L, Fransson L-A, Linker A (1979) Structure of the antithrombin-binding site in heparin. Proc Natl Acad Sci USA 76: 3198-3202

Lindsay RM, Rourke JTB, Reid BD, Linton AL, Gilchrist T, Courtney JM, Edwards RO (1977) The role of heparin on platelet retention by acrylonitrile copolymer dialysis membranes. J Lab Clin Med 89: 724-734

Lins LE, Olsson P, Hjelte MB, Larsson R, Larm O (1984) Haemodialysis in dogs with a heparin coated hollow fiber dialyser. Proc Europ Dial Transplant Assoc 21: 270-275

Lyman DJ, Klein KG, Bash JL, Fritzinger BK (1970) The interaction of platelets with polymer surfaces. Thromb Diath Haemorrh 23: 120-128

Lyman DJ, Metcalf LC, Albo D Jr, Richards KF, Lamb J (1974) The effect of chemical structure and surface properties of synthetic polymers on the coagulation of blood. III. In vivo adsorption of proteins on polymer surfaces. Trans Am Soc Artif Intern Organs 20: 474-478

Lyman DJ, Knutson K, McNeill B, Shibatani K (1975) The effects of chemical structure and surface properties on the coagulation of blood. IV. The relation between polymer morphology and protein adsorption. Trans Am Soc Artif Intern Organs 21: 49-53

Marconi W, Bartoli F, Mantovani E, Pittalis F, Settembri L, Cordova C, Musca A, Alessandri C (1979) Development of new antithrombogenic surfaces by employing platelet antiaggregating agents: preparation and characterisation. Trans Am Soc Artif Intern Organs 25: 280-285

Martin FE, Shuey HF, Saltonstall CW Jr (1970) Improved membranes for hemodialysis. J Macromol Sci- Chem A4: 635-654

Mason RG (1972) The interaction of blood hemostatic elements with artificial surfaces. Prog Hemostasis Thromb 1: 141-164

Mason RG, Read MS, Brinkhous KM (1971) Effect of fibrinogen concentration on platelet adhesion to glass. Proc Soc Exp Biol Med 137: 680-682

Mason RG, Mohammad SF, Chuang HYK, Richardson PD (1976) The adhesion of platelets to subendothelium, collagen and artificial surfaces. Semin Thromb Hemostasis 3: 98-116

Merrill EW, Salzman EW, Wong PSL, Ashford TP, Brown AH, Austen WG (1970) Polyvinyl alcohol-heparin hydrogel "G". J Appl Physiol 29: 723-730

Migonney V, Fougnot C, Josefowicz M (1988) Heparin like tubings III. Kinetics and mechanism of thrombin, antithrombin III and thrombin-antithrombin complex adsorption under controlled- flow conditions. Biomaterials 9: 413-418

Miyama H, Harumiya N, Mori Y, Tanzawa H (1977) A new antithrombogenic heparinised polymer. J Biomed Mater Res 11: 251-265

Mohammad SF, Olsen DB (1986) Reduced platelet adhesion and activation of coagulation factors on polyurethane treated with albumin-IgG complex. Trans Am Soc Artif Intern Organs 32: 323-326

Mottaghy KB, Oedekoven B, Schaich-Lester P, Pöppel K, Küpper W (1989) Application of surfaces with end point attached heparin to extracorporeal circulation with membrane lungs. Trans Am Soc Artif Intern Organs 35: 146-152

Munro MS, Eberhart RC, Maki NJ, Brink BE, Fry WJ (1983) Thromboresistant alkyl derivatised polyurethanes. Am Soc Artif Intern Organs J 6: 65-75

Murabayashi S, Nosé Y (1986) Biocompatibility: bioengineering aspects. Artif Organs 10: 114-121

Mustard JF, Packham MA (1977) Normal and abnormal haemostasis. Brit Med Bull 33: 187-192

Mustard JF, Perry DW, Ardlie NG, Packham MA (1972) Preparation of suspensions of washed platelets from humans. Brit J Haematol 22: 193-204

Neddleman SW, Hook JC (1982) Platelets and leukocytes. In: Colman RW, Hirsh J, Marder VJ, Salzman EW (eds) Hemostasis and thrombosis: basic principles and clinical practice. Lippincott, Philadelphia, USA, pp 716-725

Neumann AW, Moscarello MA, Zingg W, Hum OS, Chang SK (1979) Platelet adhesion from human blood to bare and protein coated polymer surfaces. J Pol Sci Pol Symp 66: 391-398

Niemetz J (1972) Coagulant activity of leukocytes. Tissue factor activity. J Clin Invest 51: 307-313

Nilsson L, Storm KE, Thelin S, Bagge L, Hultman J, Thorelius J, Nilsson U (1990) Heparin-coated equipment reduces complement activation during cardiopulmonary bypass in the pig. Int J Artif Organs 14: 46-48

Nosé Y (1988) Long term compatibility of artificial kidneys. Artif Organs 12:1

Nosé Y, Tajima K, Imai Y, Klain M, Mrava G, Schriber K, Urbanek K, Ogawa H (1971) Artificial heart constructed with biological material. Trans Am Soc Artif Intern Organs 17: 482-487

Ogston D (1983) The physiology of hemostasis. Croom Helm, London, England

Ohshiro T (1983) Antithrombogenic characteristics of immobilised urokinase on synthetic polymers. In: Szycher M (ed) Biocompatible polymers, metals, and composites. Technomic, Lancaster, Pennsylvania, USA pp275-299

Ohshiro T, Kosaki G (1980) Urokinase immobilised on medical polymer materials: fundamental and clinical studies. Artif Organs 4: 58-64

Packham MA, Evans G, Glynn MF, Mustard JF (1969) The effect of plasma proteins on the interaction of platelets with glass surfaces. J Lab Clin Med 73: 686-697

Paik Sung CS, Bush J, McKie DB, Merrill EW (1976) Copolymers containing aminohexyl residues in side chains. J Appl Pol Sci 20: 2603-2605

Palationos GM, Dewanjee MK, Kapadvanjwala M, Novak S, Sfakianakis GN, Kaiser GA (1990) Cardiopulmonary bypass with a surface heparinised extracorporeal perfusion system. Trans Am Soc Artif Intern Organs 36: M476-M479

Park KD, Okano T, Nojiri C, Kim SW (1988) Heparin immobilisation onto segmented polyurethaneurea- effect of hydrophilic spacers. J Biomed Mater Res 22: 977-992

Patrono C, Ciabattoni G, Pinca E, Pugliese F, Castrucci G, De Salvo A, Satta MA, Peskar BA (1980) Low dose aspirin and inhibition of thromboxane B_2 production in healthy subjects. Thromb Res 17: 317-327

Peppas NA, Merrill EW (1977) Development of semicrystalline PVA hydrogels for biomedical applications. J Biomed Mater Res 11: 423-434

Phillips DR, Jennings LK, Prasanna HR (1980) Ca^{2+}-mediated association of glycoprotein G (thrombin- sensitive protein, thrombospondin) with human blood. J Biol Chem 255: 11629-11632

Ratner BD (1981) Biomedical applications of hydrogels: review and critical appraisal. In: Williams DF (ed) Biocompatibility of clinical implant materials, vol 2. CRC Press, Boca Raton, Florida, USA, pp 145-175

Ratner BD (1982) Surface characterisation of materials for blood contact applications. In: Cooper SL, Peppas NA (eds) Biomaterials: interfacial phenomena and applications, ACS Advances in Chemistry Series, vol 199. American Chemical Society, Washington DC, USA, pp 9-23

Ratner BD (1983) Surface characterisation of biomaterials by electron spectroscopy for chemical analysis. Ann Biomed Eng 11: 313-336

Rea WJ, Whiteley D, Eberle JW (1972) Long-term membrane oxygenation without systemic heparinisation. Trans Am Soc Artif Intern Organs 18: 316-320

Richardson PD, Mohammad SF, Mason RG (1977) Flow chamber studies of platelet adhesion at controlled, spatially varied shear rates. Proc Europ Soc Artif Organs 4: 175-188

Rieger H (1980) Dependency of platelet aggregation (PA) in vitro on different shear rates. Thromb Haemostasis 44: 166

Ringoir S, Vanholder R (1986) An introduction to biocompatibility. Artif Organs 10: 20-27

Ringoir S, Vanholder R (1990) New trends in dialysis. Contrib Nephrol 82: 102-106

Saba HJ, Herion JC, Walker RI, Roberts HR (1973) The procoagulant activity of granulocytes. Proc Soc Exp Biol Med 142: 614-620

Salyer IO, Blardinelli AJ, Ball GL III, Weesner WE, Gott VL, Ramos MD (1971) New blood-compatible polymers for artificial heart applications. J Biomed Mater Res Symp 1: 105-127

Salzmann EW, Merrill EW, Binder A, Wolf CRW, Ashford TP, Austen WG (1969) Protein platelet interactions on heparinised surfaces. J Biomed Mater Res 3: 69-81

Salzman EW, Lindon J, Brier D, Merrill EW (1977) Surface-induced platelet adhesion, aggregation and release. Ann N Y Acad Sci 283: 114-127

Schmer G (1972) The biological activity of covalently immobilised heparin. Trans Am Soc Artif Intern Organs 18: 321-323

Schmer G, Teng LNL, Cole JJ, Vizzo JE, Francisco MM, Scribner BH (1976) Successful use of a totally heparin grafted hemodialysis system in sheep. Trans Am Soc Artif Intern Organs 22: 654-662

Schmer G, Teny LNL, Vizzo JE, Graefe U, Milutinovich J, Cole JJ, Scribner BH (1977) Clinical use of a totally heparin grafted hemodialysis system in uremic patients. Trans Am Soc Artif Intern Organs 23: 177-183

Schmitt E, Holtz M, Klinkmann H, Esther G, Courtney JM (1983) Heparin binding and release properties of DEAE cellulose membranes. Biomaterials 4: 309-313

Sennatore F, Bernard F, Meisner K (1986) Clinical study of urokinase-bound fibrocollagenous tubes. J Biomed Mater Res 20: 189-203

Shuman MA, Levine SP (1980) Relationship between secretion of platelet factor 4 and thrombin generation during in vitro blood clotting. J Clin Invest 65: 307-313

Stormorken (1971) Platelets, thrombosis and hemolysis. Fed Proc 30: 1551-1555

Sugitachi A, Tanaka M, Kawahara T, Takagi K (1980) Antithrombogenicity of UK-immobilised polymer surfaces. Trans Am Soc Artif Intern Organs 26: 274-278

Szycher M (1983) Thrombosis, hemostasis, and thrombolysis at prosthetic interfaces. In: Szycher M (ed) Biocompatible polymers, metals, and composites. Technomic, Lancaster, Pennsylvania, USA, pp 1-33

Tanzawa H, Mori Y, Harumiya N, Miyama H, Hori M, Ohshima N, Idezuki Y (1973) Preparation and evaluation of a new athrombogenic heparinised hydrophilic polymer for use in cardiovascular system. Trans Am Soc Artif Intern Organs 19: 188-194

Thunberg L, Bäckström G, Lindahl U (1982) Further characterisation of the antithrombin-binding sequence in heparin. Carbohydr Res 100: 393-410

Tong SD, Rolfs MR, Hsu LC (1990) Evaluation of Duraflo II heparin immobilised cardiopulmonary bypass circuits. Trans Am Soc Artif Intern Organs 36: M654-M656

Uniyal S, Brash JL, Degterev IA (1982) Influence of red blood cells and their components on protein adsorption. Am Chem Soc Adv Chem 199: 277-292

Usdin VR, Fourt L (1969) Effect of proteins on elution of heparin from anticoagulant surfaces. J Biomed Mater Res 3: 107-113

Vanholder R, Ringoir S (1989) Biocompatibility: an overview. Int J Artif Organs 12: 356-365

Videm V, Nilsson L, Venge P, Svennevig JL (1991) Reduced granulocyte activation with a heparin-coated device in an in vitro model of cardiopulmonary bypass. Artif Organs 15: 90-95

von Segesser LK, Turina M (1989) Cardiopulmonary bypass without systemic heparinisation. Performance of heparin-coated oxygenators in comparison with classic membrane and bubble oxygenators. J Thorac Cardiovasc Surg 98: 386-396

Vroman L, Adams AL, Klings M, Fischer GC, Munoz PC, Solensky RP (1972) Reactions of formed elements of blood with plasma proteins at interfaces. Ann N Y Acad Sci 283: 65-76

Vroman L, Adams AL, Fischer GC, Munoz PC (1980) Interaction of high molecular weight kininogen, factor XII and fibrinogen in plasma at interfaces. Blood 55: 156-159

Vulic I, Okano T, Kim SW, Feijen J (1988) Synthesis and characterisation of polystyrene-poly (ethylene oxide)-heparin block copolymers. J Pol Sci Pol Chem 26: 381-391

Walsh PN (1982) Platelet-coagulant protein interactions. In: Colman RW, Hirsh J, Marder VJ, Salzman EW (eds) Hemostasis and thrombosis: basic principles and clinical practice. Lippincott, Philadelphia, USA, pp 404-420

Watanabe S, Shimuzu Y, Teramatsu T, Mirachi T, Hino T (1981) The in vitro and in vivo behaviour of urokinase immobilised onto collagen-synthetic polymer composite material. J Biomed Mater Res 15: 553-563

Waugh DF, Baughman DJ (1969) Thrombin adsorption and possible relations to thrombus formation. J Biomed Mater Res 3: 145-164

Whicher SJ, Brash JL (1978) Platelet-foreign surface interactions: release of granule constituents from adherent platelets. J Biomed Mater Res 12: 181-201

Williams DF (ed) (1981) Fundamental aspects of biocompatibility. CRC Press, Boca Raton, Florida, USA

Williams DF (ed) (1987) Definitions in biomaterials. Elsevier, Amsterdam, The Netherlands

Wright DG, Kauffman JC, Terpstra GK, Graw RG, Deisseroth AB, Gallin JJ (1978) Mobilisation and exocytosis of specific (secondary) granules by human neutrophils during adherence to nylon wool infiltration leukapheresis (FL). Blood 52: 770-782

Yen SPS, Rembaum A (1971) Complexes of heparin with elastomeric positive polyelectrolytes. J Biomed Mater Res Symp 1: 83-97

Young BR, Lambrecht LK, Albrecht RM, Mosher DF, Cooper SL (1983) Platelet-protein interactions at blood-polymer interfaces in the canine test model. Trans Am Soc Artif Intern Organs 29: 442-446

Zucker MB, Vroman L (1969) Platelet adhesion by fibrinogen adsorbed onto glass. Proc Soc Exp Biol Med 131: 318-320

Zwaal RF, Comfurius P, van Deenen UM (1977) Membrane asymmetry and blood coagulation. Nature 268: 358-360

Chapter 13

Crystalline Bacterial Cell Surface Layers (S-Layers) as Combined Carrier/Adjuvants for Conjugate Vaccines

A. J. Malcolm, P. Messner, U. B. Sleytr, R. H. Smith and F. M. Unger

Introduction

The development of vaccines to prevent human and animal diseases has been a challenging task for generations of physicians, microbiologists and biochemists. Jenner's fundamental publication on inoculation of susceptible people with cowpox virus was the first documented evidence for the efficacy of vaccination (Jenner 1798). Since then, a number of vaccines against a variety of viral and bacterial infections have been used in man and animals. Other vaccines have been developed to protect against parasitic infections. More recently, experimental immunotherapies against cancers have been developed which utilise conjugate vaccine technology (for review see Bittle and Murphy 1989). Although only a few diseases have been totally eliminated through vaccination, the availability of vaccines has led to dramatic reductions in morbidity and mortality. However, there are still many infectious diseases against which vaccines need to be developed (for reviews see Bell and Torrigiani 1987; Mizrahi 1990). Moreover, problems remain regarding the efficacy of some currently available vaccines in the general population. The development of new carrier materials, capable of enhancing immunogenicity and long term immunological memory, is seen as an important contribution toward solving the remaining problems in vaccination.

In this brief overview, we report on recent work in our laboratories to develop conjugate vaccines based on S-layers. We describe the preparation and testing of conjugate vaccines using S-layers as carriers for selected, weakly immunogenic carbohydrate antigens and haptens.

Crystalline Bacterial Cell Surface Layers (S-Layers)

S-layers can be defined as two-dimensional crystalline arrays of proteinaceous subunits forming surface layers on prokaryotic cells (for reviews see Sleytr 1978; Sleytr and Messner 1983, 1988; Messner and Sleytr 1992). A detailed description of the properties of S-layers can be found in chapter 10 this volume (Pum *et al.*). Chemically, most S-layers consist of a single, homogeneous protein or glycoprotein species. The molecular masses of protein subunits have been shown to range from *ca.* 40 000 to *ca.* 220 000 by SDS-polyacrylamide gel electrophoresis. Comparison of amino acid analyses, and genetic studies on S-layers from prokaryotes of different phylogenetic origins, show that the crystalline arrays are usually composed of weakly acidic proteins. The content of hydrophobic amino acids is generally high and that of sulphur-containing amino acids is low (for reviews see Sleytr 1978; Sleytr and Messner 1983; Smit 1987; Koval 1988; Messner and Sleytr 1992).

Table 13.1. Glycan structures of eubacterial S-layer glycoproteins

---

Bacillus stearothermophilus NRS 2004/3a (Christian *et al.* 1986; Messner *et al.* 1987)

(-2αLRhap1-2αLRhap1-3ßLRhap1-)$_{n-50}$ and

(-4ßManpA2, 3(NAc)$_2$1-3αGlcpNAc1-4ßManpA2, 3(NAc)$_2$1-6αGlcp1-)$_{n-15}$

Clostridium thermohydrosulfuricum L111-69 and L110-69 (DSM 568)

(Christian *et al.* 1988)

(-4αDManp1-3αLRhap1-)$_{n-30}$

Clostridium thermohydrosulfuricum L77-66

(Altman *et al.* 1991a)

(-3αDGalpNAc1-3[αDGlcpNAc1-2ßDManp1-4]αDGalpNAc1-)$_n$

Clostridium thermosaccharolyticum D120-70

(Altman *et al.* 1990)

(-3[ßDGlcp1-6]ßDManp1-4αLRhap1-3αDGlcp1-4[αDGalp1-2]αLRhap1-)$_n$ and

(-4ßGlcpNAc1-3[(αGalp)$_{0.5}$1-4]ßDManpNAc1-)$_n$

Bacillus alvei CCM 2051

(Altman *et al.* 1991b)

(-3ßDGalp1-4[αDGlcp1-6]ßDManpNAc1-)$_{n-40}$

Clostridium symbiosum HB25

(Messner *et al.* 1990)

(-6αDManpNAc1-4ßDGalpNAc1-3αDBacpNAc1-4αDGalpNAc1-PO$_3$H-)$_{n-15}$

Abbreviations: D-Glcp, D-Glucopyranose; Gal, galactose; Man, mannose; Rha, rhamnose; GlcNAc, GalNAc, ManNAc, corresponding *N*-acetylamino- sugars; ManpA2,3(NAc)$_2$, 2,3-diacetamido-2,3-dideoxy-mannuronic acid; D-BacpNAc, N-acetyl-D-bacillosamine (2-acetamido-4-amino-2,4,6-trideoxy- D-glucose)

A remarkable feature of most archaebacteria (Kandler 1982; König 1988; Nusser *et al.* 1988; Lechner and Wieland 1989) and some eubacteria (Küpcü *et al.* 1984; Messner and Sleytr 1988a, 1991) is their ability to produce glycosylated S-layer proteins. Several primary structures of glycan chains of S-layer glycoproteins have been studied by controlled degradation reactions, mass spectroscopy, and ^{1}H and ^{13}C nuclear magnetic resonance techniques. As summarised in Table 13.1, almost all glycan chains characterised to date are polymers of linear or branched repeating sequences (two to six monosaccharide units) consisting of hexoses, deoxy sugars, amino sugars and uronic acids, with some repeating units containing sulphate or phosphate residues (for review see Messner and Sleytr 1991). They strongly resemble bacterial O antigenic polysaccharides and are thus significantly different from eukaryotic glycoprotein glycans (Kornfeld and Kornfeld 1980). Their location on the cell surface has been demonstrated by electron microscopy using topographical marker molecules and chemical modification of the carbohydrates (Sára *et al.* 1989; see also Pum *et al.* chapter 10 this volume).

Carbohydrates as Antigens or Haptens

Heidelberger and Avery (1923) demonstrated that the type specific antigens of pneumococci are polysaccharides. Bacterial capsular polysaccharides are cell surface antigens composed of identical repeat units which form extended saccharide chains. Such antigenic exopolysaccharide structures are present on pathogenic bacteria and have been identified on *Escherichia coli, Neisseria meningitidis, Haemophilus influenzae*, Group B *Streptococcus, Streptococcus pneumoniae* and other species (for review see Kenne and Lindberg 1983).

Examples of mammalian cell surface carbohydrates are specific blood group determinants and "tumour-associated" antigens. Oncogenically transformed cells often display structures and profiles of cell surface carbohydrates distinctly different from those of non-transformed cells. In both types of example, the glycans consist of only a few monosaccharides (for review see Hakomori and Kannagi 1986). The glycan structures by themselves are usually not antigenic, but constitute haptens in conjunction with protein or glycoprotein matrices.

A general feature of saccharide antigens is their inability to elicit a T-cell mediated immune response. They are therefore considered thymus-independent antigens. Conjugation of the polysaccharide antigens or of immunologically non- reactive carbohydrate haptens to thymus-dependent antigens, e.g. proteins, enhances their immunogenicity considerably. The protein stimulates carrier-specific T-helper cells which play a role in the induction of anti-carbohydrate antibody synthesis (for review see Bixler and Pillai 1989).

Conjugate Vaccines

Avery and Goebel were the first to prepare conjugate vaccines against bacterial infections (Avery and Goebel 1929; Goebel and Avery 1929). More recently, Schneerson and co-workers (1980), Anderson (1983), Gordon (1984) and other investigators have developed conjugate vaccines against *Haemophilus influenzae* type b (Hib) infections based on polyribitol-phosphate (PRP) linked to tetanus toxoid. The principles underlying the preparation of Hib vaccines have been extended to other bacterial polysaccharides and to peptides of bacterial, viral and parasitic origin (for reviews see Cruse and Lewis 1989; Dick and Burret 1989). Tetanus and diptheria toxoids have been widely used as carriers for conjugate vaccines. For experimental purposes, other carriers such as bovine serum albumin (BSA), keyhole limpet or *Limulus polyphemus* hemocyanins, thyroglobulin, ß-galactosidase, immunoglobulin G (IgG), synthetic polymers of α-amino acids, as well as human erythrocytes, liposomes, and silica beads have been used (for reviews see Sela 1987; Altman and Dixon 1989).

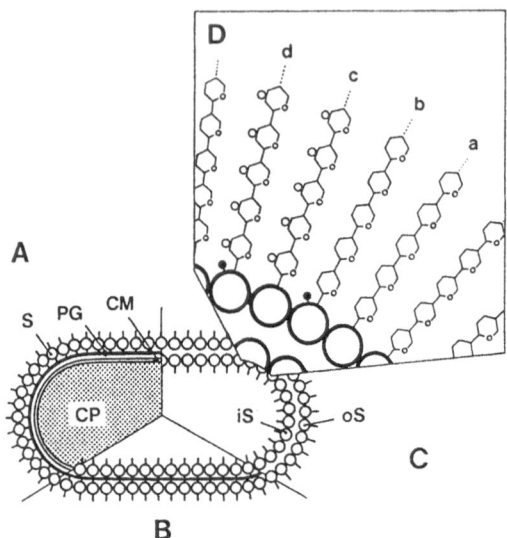

Fig. 13.1. Preparation scheme for glutaraldehyde-fixed S-layer conjugates. (A) Intact bacterial cell; S, S- layer; PG, peptidoglycan; CM, cytoplasmic membrane; CP, cytoplasm. (B) After removing the cell content and the cytoplasmic membrane an additional S-layer assembles on the inside of the peptidoglycan layer. (C) Double S-layer fragments are obtained by digesting the peptidoglycan layer with lysozyme. The two S-layers are arranged in a mirror-symmetric fashion. Functional groups of the protein and carbohydrate moieties, exposed on the inner (iS) and outer (oS) S-layer surface are available for binding ligands. (D) Schematic illustration of the possibilities for covalent binding of haptens. (a) Non-activated glutaraldehyde-fixed S-layer glycoprotein. The ligands (●, O) can be linked either to the protein moiety (b) or to the glycan chains (c) or to both of them (d).

Preparation of Carbohydrate S-Layer Conjugates

In conjugate vaccines, the antigens or haptens are bound to a protein by covalent linkages. Usually the protein molecules are present as monomers in solution or dispersed as unstructured aggregates. With traditional carriers, reproducible attachment of ligands to carrier proteins is often difficult to achieve.

Due to the crystalline nature of S-layer (glyco)proteins, the amino, carboxyl, or hydroxyl groups available for hapten binding occur on each protomer in identical positions and orientations (Pum *et al.* chapter 10 this volume). Ligands can be immobilised onto these precisely defined matrices (Sleytr *et al.* 1987; Sára and Sleytr 1989; Sleytr *et al.* 1991). In our experiments, we used the S-layer glycoproteins isolated from *Bacillus stearothermophilus* NRS 2004/3a (henceforth abbreviated as 3a), Bacillus alvei CCM 2051 (2051), *Clostridium thermohydrosulfuricum* L111-69 (L111), and *Clostridium thermosaccharolyticum* D120-70 (D120) (see Table 13.1) and the non-glycosylated S-layer protein from *Bacillus stearothermophilus* PV72 (PV72).

Two forms of S-layer were prepared for this study, namely, glutaraldehyde-fixed double S-layer sacculi or S-layer self-assembly products. To prepare the sacculi, whole bacterial cell preparations (Fig. 13.1A) were opened by gentle sonication for releasing the cell content. After treatment with Triton X-100 to remove the cytoplasmic membrane, a second S-layer was formed from excess S-layer material on the inner leaflet of the peptidoglycan (Fig. 13.1B). Following extensive washing of the purified cell walls, glutaraldehyde was used to fix the double S-layer sacculi. These have mostly shape and external appearance of intact bacterial cells (Fig. 13.1C; Sára and Sleytr, 1989). Self-assembly products were generated after extraction of the S-layer material from purified cell wall preparations by treatment with chaotropic agents (e.g. guanidine hydrochloride). These agents were subsequently removed by dialysis (Messner and Sleytr 1988b). Glutaraldehyde-fixed or unfixed self-assembly products were then used as the immobilisation matrix (P. Messner, M.A. Mazid, F.M. Unger, U.B. Sleytr, submitted for publication).

A variety of carbohydrate haptens were immobilised, including the blood group A-trisaccharide (αGalNAc1-3[αFuc1-2]ßGal) and the "tumour associated" T (ßGal1-3αGalNAc) and Y antigens (αFuc1-2ßGal1-4[αFuc1-3]ßGalNAc). These synthetic oligosaccharide haptens were used as their 8-methoxycarbonyloctyl glycosides (for review see Lemieux 1987). Oligosaccharides obtained by hydrolysis of polysaccharides from *Streptococcus pneumoniae* were also used in this study. Different chemical reactions were utilised for the covalent attachment of the ligands (Table 13.2). With S-layer glycoproteins, ligands can be coupled to the protein portion, or to the glycan chains (Sára and Sleytr 1989; Messner *et al.* 1991). This approach offers a way for preparing multivalent conjugates (Fig. 13.1D). Periodate, epichlorohydrin, or divinyl sulfone can be used to activate the carbohydrate chains of the S-layer glycoproteins. The desired antigens or haptens can be attached to the modified glycan chains through Schiff base formation, nucleophilic opening of oxiran rings, or addition of polar groups to activated vinyl groups (P. Messner, M.A. Mazid, F.M. Unger, U.B. Sleytr, submitted for publication; B. Sefcik, unpublished results). Alternatively, antigens or haptens can be coupled to the protein portion of the S-layer following activation of carboxyl groups with 1-ethyl- 3,3'-dimethyl-(aminopropyl)-carbodiimide (EDC) or via amino groups which are converted to sulfhydryl groups (R. Tomasits, unpublished results) (Table 13.2).

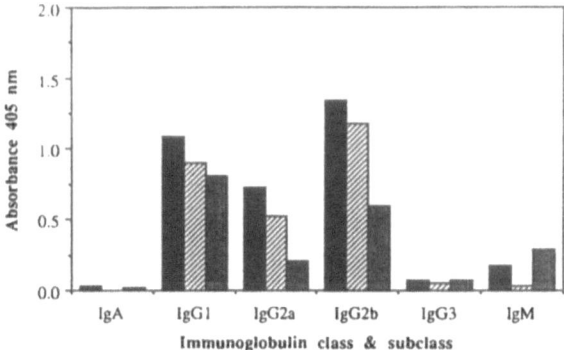

Fig. 13.2. ELISA immunoglobulin typing of S-layer antisera. Tertiary antibody response of female Balb/c mice immunised intraperitoneally with 20 µg of unfixed PV72 (■) 2051 (▨) or L111 (▨) S-layers, isotyped at 10^{-3} serum dilution by enzyme linked immunosorbent assay (ELISA).

Table 13.2. Reaction schemes of the applied immobilisation methods

a. Periodate activation (Reductive amination)

S-layer glycan chain

$+$ $NaIO_4$ $\longrightarrow$ Sodium periodate   activated material $+$ ligand

$NaBH_4$ or $NaCNBH_3$ $\longrightarrow$ Schiff's base   immobilised ligand

b. Divinyl sulfone activation

S-layer glycan chain $+$ $H_2C=CH-S-CH=CH_2$ (Divinyl sulfone) $\longrightarrow$ activated material

$+$ ligand $\longrightarrow$ immobilised ligand

c. Epichlorohydrin activation

S-layer glycan chain $+$ $Cl-CH_2-CH-CH_2$ (Epichlorohydrin) $\longrightarrow$ activated material

$+$ ligand $\longrightarrow$ immobilised ligand

d. Carbodiimide activation

S-layer protein $+$ 1-ethyl-3,3'-dimethyl-(aminopropyl)-carbodiimide (EDC) $\longrightarrow$ activated material

$+$ ligand $\longrightarrow$ immobilised ligand

or

acidic sugar $+$ EDC $\longrightarrow$ activated material

S-layer protein $+$ $\longrightarrow$ immobilised ligand

e. Iminothiolane / 2,2'-Dithiopyridine activation

S-layer protein $+$ Iminothiolane $\longrightarrow$ amino groups converted to SH-groups and activated

$+$ 2,2'-dithiopyridine $\longrightarrow$ immobilised ligand

a

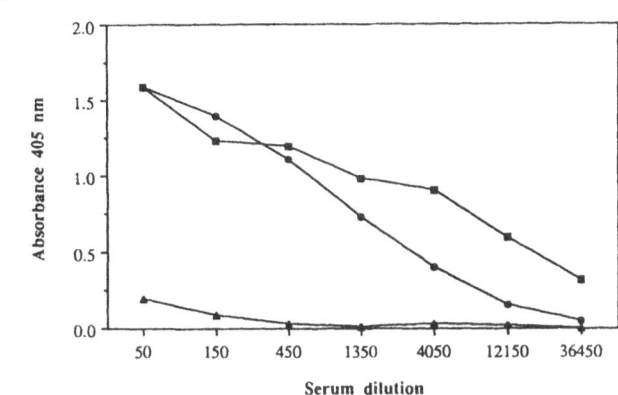

b

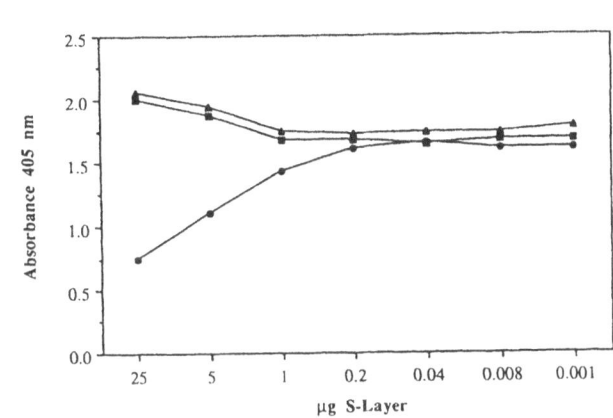

Fig. 13.3. Antibody titer and inhibition ELISA of anti-3a S-layer antiserum. (a) ELISA titer of tertiary antibody response to 3a S-layer (●), to 3a S-layer with FCA (■), and FCA (▲) as a negative control. (b) An inhibition ELISA with secondary 3a S-layer antiserum at 4×10^{-3} dilution with 3a (●), L111 (■), and D120 (▲) S-layers as soluble inhibitory antigens.

Antibody Response to S-Layer Conjugates

The ability of S-layer conjugates to elicit antibodies to haptens of weak immunogenicity has been examined. A variety of types and forms of S-layers and different immunisation schedules has been studied to establish the best procedure for eliciting antibodies.

Intraperitoneal immunisation of mice with unfixed glycosylated 3a, L111, D120 and 2051 S-layers (see Table 13.1) or the non-glycosylated unfixed PV72 S-layer, elicited anamnestic and long lasting antibody responses. Figure 13.2 illustrates the immunoglobulin class and subclasses of antisera elicited after tertiary immunisation with unfixed PV72, 2051 and L111 S-layers.

Unfixed S-layers elicit good antibody responses without the necessity of an extraneous adjuvant. Figure 13.3a demonstrates the similarity of the antibody response to the unfixed 3a S-layers with or without Freund's adjuvant. Fixed L111 S-layer also elicited an antibody response, although the response was tenfold less than the response to unfixed L111. Immunisation with adjuvants (Freund's Complete Adjuvant (FCA), Freund's Incomplete Adjuvant (FIA), dimethyl-dioctadecyl ammonium bromide (DDA), 2,6,10,14-tetramethylpentadecane (pristane) or Adjuvax®) enhanced the response to fixed L111.

Antibodies elicited to S-layers from strains 3a, D120, 2051, L111 and PV72 did not cross react, as determined by an inhibition ELISA (Malcolm *et al.* 1984; A.J. Malcolm *et al.*, submitted for publication). An example is shown in Fig. 13.3b. The binding of anti-3a S-layer antibodies to a 3a antigen-coated ELISA plate was inhibited by free 3a antigen; free L111 or D120 antigen did not inhibit binding to the ELISA plate.

An initial, limited study was undertaken to examine the antibody response to the "tumour-associated" T (ßGal1-3αGalNAc) and Y antigens (αFuc1-2ßGal1-4[αFuc1-3]ßGalNAc) coupled to S-layers (as outlined in "Preparation of Carbohydrate S-Layer Conjugates"). Specific antibodies to these structures have been elicited and a class switch from IgM to IgG was observed. The amount of hapten coupled to the S-layer carrier was important in eliciting an antibody response. Higher molar ratios of hapten to S-layer induced better responses.

Recent efforts have been focused on the development of S-layer conjugates which are capable of eliciting immunoprotective antibodies to capsular polysaccharides of *Streptococcus pneumoniae*. For this prototype vaccine, we selected three serotypes (3, 6B and 8) which are associated with bacteremic disease and high case fatality rates.

In initial studies, mice were immunised intraperitoneally with *S. pneumoniae* type 8 capsular *polysaccharide* (CPS) alone or coupled to unfixed S-layers. The antibody responses elicited were typical of T-independent antigens. The tertiary antibody response to serotype 8 polysaccharide S-layer conjugates (PV72, 2051, L111) was mainly of the IgM and IgA classes, with very low levels of IgG isotypes. Results from bactericidal assays (used to quantify levels of immunoprotective antibodies) demonstrate that no immunoprotective antibodies were elicited by injection of polysaccharide S-layer conjugates or the polysaccharides alone.

With these essentially negative results, subsequent efforts were directed to developing immunogenic *oligosaccharide* S-layer conjugates. Procedures were established to prepare and isolate oligosaccharides from serotypes 3, 6B and 8. Type 3 and type 8 oligosaccharides have been coupled to unfixed S-layers and the conjugates injected intraperitoneally into mice. Figure 13.4 shows the isotypic class switch (IgM to IgG response) when 8-oligosaccharide-2051 S-layer conjugates (8-2051) were injected. These conjugates appear to possess intrinsic adjuvant properties; no additional adjuvant material was necessary to elicit specific antibodies or induce class switching.

Immunoprotection *in vitro* was observed with antibodies to type 8 oligosaccharides when coupled to unfixed PV72, 2051, or L111 S-layers. Cultures of *Streptococcus pneumoniae* serotype 8 were mixed with tertiary antisera to 8- oligosaccharide 2051, PV72 and L111 S-layer conjugates and to serotype 8 CPS, and then streaked onto blood agar plates to assay for surviving bacteria. Antisera to 8-CPS did not reduce growth of *S. pneumoniae*, whereas antisera to 8- oligosaccharide S-layer conjugates completely inhibited growth. In other studies, oral/nasal administration of type 8 oligosaccharide-2051 S-layer conjugates elicited moderate titre increases of protective antibodies against *S. pneumoniae* type 8. Also, rabbits immunised parenterally with *S. pneumoniae* type 8 oligosaccharide S- layer conjugates produced immunoprotective

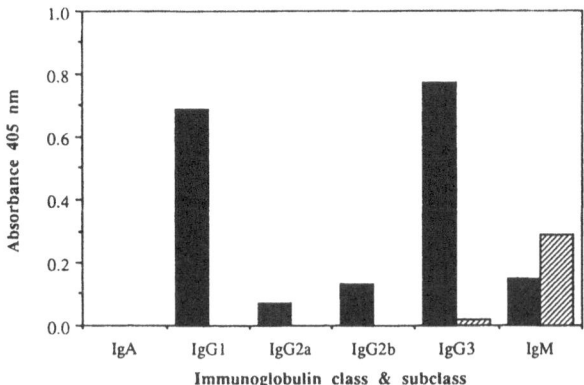

Fig. 13.4. ELISA immunoglobulin typing of antisera to a 8-oligosaccharide S-layer conjugate. Tertiary antibody responses to a 8-2051 conjugate (■) and to capsular polysaccharide serotype 8 (▨), isotyped at 4×10^{-2} serum dilution by ELISA.

antibodies. Studies are continuing with conjugates which contain oligosaccharides corresponding to different numbers of repeat units (2-4, 4-6, or 8-12 repeat units) to determine which conjugate elicits the best immunoprotective antibody response.

T Cell Responses to S-Layer Conjugates

To investigate the carrier/adjuvant potential of S-layers with respect to immunotherapy of cancers, glutaraldehyde-fixed S-layer conjugates containing the "tumour-associated" T-disaccharide were prepared as outlined in "Preparation of Carbohydrate S-Layer Conjugates". Mice were immunised intramuscularly with these conjugates as described by Smith and Ziola (1986). Delayed-type hypersensitivity (DTH) responses were measured as footpad swelling seven days later, following a footpad challenge with T-disaccharide conjugated to a different glutaraldehyde-fixed S-layer. Initial experiments were performed to examine the optimal conditions for induction of a specific DTH response (R.H. Smith, P. Messner, L.R. Lamontagne, U.B. Sleytr, F.M. Unger, submitted for publication). The greatest DTH response was elicited when mice were immunised with 10 μg of fixed T-disaccharide-3a conjugates; an immunisation dose of 5 μg was necessary to prime for DTH response. A challenge dose of 10 μg of T-S-layer conjugate per mouse was found to be optimal. To examine the efficacy of S-layers as carrier/adjuvants, mice were immunised with either haptenated S-layers, bovine serum albumin (BSA) conjugates, or BSA conjugates precipitated with aluminium hydroxide $(Al(OH)_3$; Fig. 13.5).

Mice immunised with T-3a and challenged with T-L111 S-layer preparations displayed the strongest DTH responses. No significant DTH response was observed with mice immunised and challenged with T-BSA conjugates. Non-immunised mice challenged with T-L111 or BSA were negative controls for this experiment. Precipitation of T-BSA conjugates with aluminium hydroxide resulted in increased responses. Oral/nasal administration of T-S-layer conjugates also resulted in strong, hapten-specific DTH responses. In another experiment, lymphocytes from mice primed

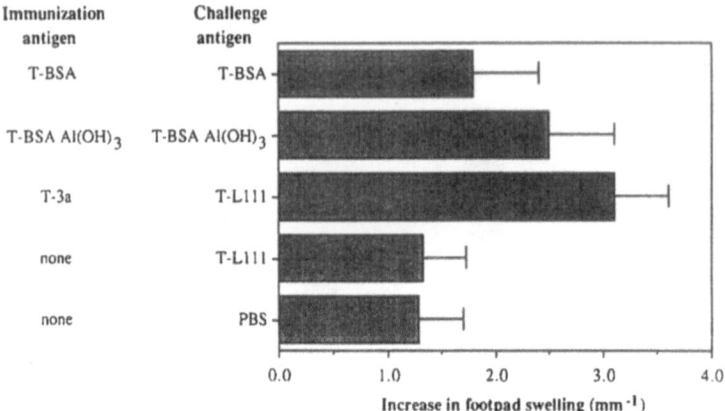

Fig. 13.5. Delayed-type hypersensitivity responses generated by glutaraldehyde-fixed S-layer conjugates. Female Balb/c mice were immunised intramuscularly with T-3a S-layer, T-BSA or T-BSA conjugates precipitated with aluminium hydroxide. Seven days later, mice were footpad challenged with T-L111 conjugates, T-BSA or T-BSA with aluminium hydroxide. Footpad swelling was measured 24 h later (mean swelling ± standard deviation).

with haptenated S-layers were stimulated *in vitro*, and then adoptively transferred to naive mice. Footpad-challenge of these mice with the same hapten bound to a different S-layer resulted in a strong DTH response against the respective hapten. The *in vitro* culture step was also performed in the presence of monoclonal antibodies directed against specific T-cell markers. The DTH response was abrogated with anti-helper T cell antibody, indicating that the hapten-specific DTH responses were mediated by helper T cells (R.H. Smith, P. Messner, L.R. Lamontagne, U.B. Sleytr, F.M. Unger, submitted for publication). Results from this study also show that the glutaraldehyde-fixed S-layers themselves (3a, L111, and 2051) do not cross-react at the level of helper T cells.

Summary

S-layers are two-dimensional crystalline arrays of proteinaceous subunits forming surface layers on prokaryotic cells. As a strain specific characteristic the S-layer protein subunits can either occur glycosylated or non-glycosylated. The S-layers have been used as prototype vaccine carriers mainly in two types of preparation: as glutaraldehyde-fixed, double S-layer sacculi or as unfixed S-layer self-assembly products. Antigens or haptens can be covalently linked to either the protein or the glycan portions of S-layer glycoproteins, or both. In this chapter, we present evidence that S-layers are effective carrier/adjuvants for eliciting T and B cell responses to specific antigens.

Unfixed S-layer self-assembly products containing covalently attached oligosaccharides isolated from hydrolysis mixtures of *Streptococcus pneumoniae* type 8 capsules elicited excellent titres of protective antibodies against *S. pneumoniae* type 8 with parenteral immunisation regimens. Effective class switching from IgM to IgG subclasses was observed upon secondary and tertiary immunisation. Oral/nasal administration of S-layer conjugates induced moderate titre increases of protective antibodies against *S. pneumoniae* type 8. Glutaraldehyde-fixed S-layer conjugates containing synthetic, tumour-associated oligosaccharides such as the T and Y structures were shown to prime for hapten- specific delayed-type hypersensitivity responses in mice upon challenge seven days after primary intramuscular immunisation. Strong increases in hapten-specific DTH responses were also elicited by oral/nasal immunisation. The results of adoptive transfer experiments indicate that the DTH responses are mediated by T-helper lymphocytes specific for the tumour associated haptens.

The S-layers investigated are not immunologically cross-reactive. Secondary and tertiary immunisations may therefore be performed using the same hapten bound to different S-layers. Thus, the hapten-specific responses can be preferentially amplified. Currently, tetanus and diptheria toxoids are widely used as vaccine carriers. These toxoids are also used for immunisation of infants, and tetanus toxoid is often administered as a prophylactic measure following trauma. Such repeated vaccination may cause a state of tolerance, effectively disabling the response to any hapten bound to these toxoid carriers. We believe that the carrier-suppression phenomena observed with toxoid proteins could be avoided by using different S-layers. Immunologically distinct S-layers could be used as carriers for each vaccine.

The data gathered to date indicate that S-layers have good potential as carrier/adjuvants. They are capable of eliciting immunoprotective antibody and T- helper cell responses against small carbohydrate haptens. The results of a comparative DTH study involving S-layer conjugates in parallel with bovine serum albumin/aluminium hydroxide combinations and the finding that unfixed S-layers elicit immunoprotective antibodies, indicate that the aggregate nature of the S-layer vaccines endows them with an intrinsic adjuvant property. The addition of extraneous adjuvant is therefore not necessary for effective vaccination with S-layer conjugates.

Acknowledgements

This work was supported in part by the "Fonds zur Förderung der wissenschaftlichen Forschung in Österreich" project P7757 and the "Österreichisches Bundesministerium für Wissenschaft und Forschung".

The S-layer research at Chembiomed Ltd. was partially funded by the Canadian Bacterial Diseases Network, one of the Networks of Centres of Excellence of Canada.

References

Altman A, Dixon FJ (1989) Immunomodifiers in vaccines. In: Bittle JL, Murphy FL (eds) Vaccine biotechnology. Advances in veterinary science and comparative medicine, vol 33. Academic Press, San Diego, pp 301-343

Altman E, Brisson J-R, Messner P, Sleytr UB (1990) Chemical characterisation of the regularly arranged surface layer glycoprotein of *Clostridium thermosaccharolyticum* D120-70. Eur J Biochem 188:73-82

Altman E, Brisson J-R, Gagné S, Kolbe J, Messner P, Sleytr UB (1991a) Structure of the glycan chain from the surface layer glycoprotein of *Clostridium thermohydrosulfuricum* L77-66. Glycoconjugate J 8:242

Altman E, Brisson J-R, Messner P, Sleytr UB (1991b) Chemical characterisation of the regularly arranged surface layer glycoprotein of *Bacillus alvei* CCM 2051. Biochem Cell Biol 69:72-78

Anderson P (1983) Antibody responses to *H. influenzae* type b and diphtheria toxin induced by conjugates of oligosaccharides of the type b capsule with the nontoxic protein CRM197. Infect Immun 39:233-238

Avery OT, Goebel WF (1929) Chemo-immunological studies on conjugated carbohydrate proteins. II Immunological specificity of synthetic sugar-protein antigens. J Exp Med 50:533-550

Bell R, Torrigiani G (eds) (1987) Towards better carbohydrate vaccines. J Wiley and Sons, Chichester

Bittle JL, Murphy FL (eds) (1989) Vaccine biotechnology. Advances in veterinary science and comparative medicine, vol 33. Academic Press, San Diego

Bixler GS Jr, Pillai S (1989) The cellular basis of the immune response to conjugate vaccines. In: Cruse JM, Lewis RE Jr (eds) Conjugate vaccines. Contributions to microbiology and immunology, vol 10. Karger, Basel, pp 18-47

Christian R, Schulz G, Unger FM, Messner P, Küpcü Z, Sleytr UB (1986) Structure of a rhamnan from the surface layer glycoprotein of *Bacillus stearothermophilus* strain NRS 2004/3a. Carbohydr Res 150:265-272

Christian R, Messner P, Weiner C, Sleytr UB, Schulz G (1988) Structure of a glycan form the surface-layer glycoprotein of *Clostridium thermohydrosulfuricum* strain L111-69. Carbohydr Res 176:160-163

Cruse JM, Lewis RE Jr (1989) Contemporary trends in conjugate vaccine development. In: Cruse JM, Lewis RE Jr (eds) Conjugate vaccines. Contributions to microbiology and immunology, vol 10. Karger, Basel, pp 1-10

Dick WE, Burret M (1989) Glycoconjugates of bacterial carbohydrate antigens. A survey and consideration of design and preparation factors. In: Cruse JM, Lewis RE Jr (eds) Conjugate vaccines. Contributions to microbiology and immunology, vol 10. Karger, Basel, pp 48-114

Goebel WF, Avery OT (1929) Chemo-immunological studies on conjugated carbohydrate-proteins. I. The synthesis of p-aminophenol ß-glucoside, p-aminophenol ß-galactoside, and their coupling with serum globulin. J Exp Med 50:521-531

Gordon LK (1984) Characterisation of a hapten-carrier conjugate vaccine: H. *influenzae*-diphtheria conjugate vaccine. In: Chanock RM, Lerner RA (eds) Vaccines '84. Modern approaches to vaccines. Cold Spring Harbor Laboratory, Cold Spring Harbor, New York, pp 393-396

Hakomori S, Kannagi R (1986) Carbohydrate antigens in higher animals. In: Weir DM, Herzenberg LA, Blackwell C, Herzenberg LA (eds) Handbook of experimental immunology, 4th ed, vol 1, Immunochemistry. Blackwell, Oxford, pp 9.1-9.39

Heidelberger M, Avery OT (1923) The soluble specific substance of Pneumococcus I. J Exp Med 38:73-79

Jenner E (1798) An inquiry into the uses and effects of the variolea vaccine. Sampson Low, London

Kandler O (1982) Cell wall structures and their phylogenic implications. Zbl Bakt Hyg I Abt Orig C 3:149-160

Kenne L, Lindberg B (1983) Bacterial polysaccharides. In: Aspinall GO (ed) The Polysaccharides, vol 2. Academic Press, New York, pp 287-363

König H (1988) Archaeobacterial cell envelopes. Can J Microbiol 34:395-406

Kornfeld R, Kornfeld S (1980) Structure of glycoproteins and their oligosaccharide units. In: Lennarz WJ (ed) The biochemistry of glycoproteins and proteoglycans. Plenum Press, New York, pp 1-34

Koval SF (1988) Paracrystalline protein surface arrays on bacteria. Can J Microbiol 34:407-414

Küpcü Z, MÑrz L, Messner P, Sleytr UB (1984) Evidence for the glycoprotein nature of the crystalline cell wall surface layer of *Bacillus stearothermophilus* strain NRS 2004/3a. FEBS Lett 173:185-190

Lechner J, Wieland F (1989) Structure and biosynthesis of prokaryotic glycoproteins. Annu Rev Biochem 58:173-194

Lemieux RU (1987) Applications of synthetic complex oligosaccharides to areas of molecular biology. In: Bell R, Torrigiani G (eds) Towards better carbohydrate vaccines. J Wiley and Sons, Chichester, pp 41-58

Malcolm AJ, Shipman RC, Levy JG (1984) A monoclonal antibody to myelogenous leukemia: isolation and characterisation. Exper Haematology 12:539-547

Messner P, Sleytr UB (1988a) Asparaginyl-rhamnose: a novel type of protein-carbohydrate linkage in an eubacterial surface-layer glycoprotein. FEBS Lett 228:317-320

Messner P, Sleytr UB (1988b) Isolation and purification of S-layers from Gram-positive and Gram-negative bacteria. In: Hancock IC, Poxton IR (eds) Bacterial cell-surface techniques. J Wiley and Sons, Chichester, pp 97-104

Messner P, Sleytr UB (1991) Bacterial surface layer glycoproteins. Glycobiology 1:545-551

Messner P, Sleytr UB (1992) Crystalline bacterial cell-surface layers. In: Rose AH, Tempest DW (eds) Advances in microbial physiology, vol 33. Academic Press, London, in press

Messner P, Sleytr UB, Christian R, Schulz G, Unger FM (1987) Isolation and structure determination of a diacetamidodideoxyuronic acid-containing glycan from the S-layer glycoprotein of *Bacillus stearothermophilus* NRS 2004/3a. Carbohydr Res 168:211-218

Messner P, Bock K, Christian R, Schulz G, Sleytr UB (1990) Characterisation of the surface layer glycoprotein of *Clostridium symbiosum* HB25. J Bacteriol 172:2576-2583

Messner P, Küpcü S, Sára M, Pum D, Sleytr UB (1991) Characterisation and biotechnological application of eubacterial glycoproteins. In: Conradt HS (ed) Protein glycosylation: cellular, biotechnological and analytical Aspects. GBF Monographs, vol 15. VCH Publishers, Weinheim, pp 111-116

Mizrahi A (ed) (1990) Bacterial vaccines. Advances in biotechnological processes, vol 13. Wiley-Liss, New York

Nusser E, Hartmann E, Allmeier H, König H, Paul G, Stetter KO (1988) A glycoprotein surface layer covers the pseudomurein sacculus of the extreme thermophile *Methanothermus fervidus*. In: Sleytr UB, Messner P, Pum D, Sára M (eds) Crystalline bacterial cell-surface layers. Springer, Berlin pp 21-25

Sára M, Sleytr UB (1989) Use of regularly structured bacterial cell envelope layers as matrix for the immobilisation of macromolecules. Appl Microbiol Biotechnol 30:184-189

Sára M, Küpcü S, Sleytr UB (1989) Localisation of the carbohydrate residue of the S-layer glycoprotein from *Clostridium thermohydrosulfuricum* L111-69. Arch Microbiol 151:416-420

Schneerson R, Barrera O, Sutton A, Robbins JB (1980) Preparation, characterisation and immunogenicity of *H. influenzae* type B polysaccharide-protein conjugates. J Exp Med 152:361-376

Sela M (1987) The choice of carrier. In: Arnon R (ed) Synthetic vaccines, vol I. CRC Press, Boca Raton, Fla, pp 83-92

Sleytr UB (1978) Regular arrays of macromolecules on bacterial cell walls: structure, chemistry, assembly, and function. Int Rev Cytol 53:1-64

Sleytr UB, Messner P (1983) Crystalline surface layers on bacteria. Annu Rev Microbiol 37:311-339

Sleytr UB, Messner P (1988) Crystalline surface layers on procaryotes. J Bacteriol 170:2891-2897

Sleytr UB, Mundt W, Messner P (1987) Pharmazeutische Struktur. German patent application, No P 37 27 987.4, 21.8.1987

Sleytr UB, Mundt W, Messner P, Smith RH, Unger FM (1991) Immunogenic compositions containing ordered carriers. US patent, No 5,043,158

Smit J (1987) Protein surface layers of bacteria. In: Inouye M (ed) Bacterial outer membranes as model systems. J Wiley and Sons, New York, pp 343-376

Smith RH, Ziola B (1986) Cyclophosphamide and dimethyl-dioctadecyl ammonium bromide immunopotentiate the delayed-type hypersensitivity response to inactivated enveloped viruses. Immunology 58:245-250

Subject Index